Geometric Measure Theory
A Beginner's Guide
Fifth Edition

Here as a child I watched my mom blow soap bubbles. My dad also encouraged all my interests. This book is dedicated to them with admiration.

Photograph courtesy of the Morgan family; taken by the author's grandfather, Dr. Charles W. Selemeyer

Geometric Measure Theory
A Beginner's Guide

Fifth Edition

Frank Morgan

Department of Mathematics and Statistics
Williams College Williamstown, Massachusetts

Illustrated by
James F. Bredt
Co-Founder, Z Corporation

AMSTERDAM • BOSTON • HEIDELBERG • LONDON
NEW YORK • OXFORD • PARIS • SAN DIEGO
SAN FRANCISCO • SINGAPORE • SYDNEY • TOKYO
Academic Press is an imprint of Elsevier

Academic Press is an imprint of Elsevier
125 London Wall, London EC2Y 5AS, UK
525 B Street, Suite 1800, San Diego, CA 92101-4495, USA
50 Hampshire Street, 5th Floor, Cambridge, MA 02139, USA
The Boulevard, Langford Lane, Kidlington, Oxford OX5 1GB, UK

Chapter headings: John M. Sullivan's computer rendering of a soap bubble cluster obtained
by area minimization on Ken Brakke's Evolver. See Preface, Chapter 13, and Section 16.8.
© John M. Sullivan. All rights reserved.

Back cover: The author holds a student-commissioned stained glass window phase portrait of double bubbles in the
3D torus. See Section 19.7.

Notices
Knowledge and best practice in this field are constantly changing. As new research and experience broaden our
understanding, changes in research methods, professional practices, or medical treatment may become necessary.

Practitioners and researchers must always rely on their own experience and knowledge in evaluating and using any
information, methods, compounds, or experiments described herein. In using such information or methods they
should be mindful of their own safety and the safety of others, including parties for whom they have a professional
responsibility.

To the fullest extent of the law, neither the Publisher nor the authors, contributors, or editors, assume any liability
for any injury and/or damage to persons or property as a matter of products liability, negligence or otherwise, or
from any use or operation of any methods, products, instructions, or ideas contained in the material herein.

Library of Congress Cataloging-in-Publication Data
A catalog record for this book is available from the Library of Congress

British Library Cataloguing-in-Publication Data
A catalogue record for this book is available from the British Library.

ISBN: 978-0-12-804489-6

For information on all Academic Press publications
visit our website at http://www.elsevier.com/

Working together
to grow libraries in
developing countries

www.elsevier.com • www.bookaid.org

Cover image: Photograph courtesy of the Morgan family; taken by the author's grandfather, Dr. Charles
W. Selemeyer.

Publisher: Nikki Levi
Acquisition Editor: Graham Nisbet
Editorial Project Manager: Susan Ikeda
Production Project Manager: Poulouse Joseph
Designer: Mark Rogers

Contents

Part II: Applications 119

Preface

Singular geometry governs the physical universe: soap bubble clusters meeting along singular curves, black holes, defects in materials, chaotic turbulence, crystal growth. The governing principle is often some kind of energy minimization. Geometric measure theory provides a general framework for understanding such minimal shapes, a priori allowing any imaginable singularity and then proving that only certain kinds of structures occur.

Jean Taylor used new tools of geometric measure theory to derive the singular structure of soap bubble clusters and sea creatures, recorded by J. Plateau over a century ago (see Section 13.9). R. Schoen and S.-T. Yau used minimal surfaces in their original proof of the positive mass conjecture in cosmology, extended to a proof of the Riemannian Penrose Conjecture by H. Bray. David Hoffman and his collaborators used modern computer technology to discover some of the first new complete embedded minimal surfaces in a hundred years (Figure 6.1.3), some of which look just like certain polymers. Other mathematicians are now investigating singular dynamics, such as crystal growth. New software computes crystals growing amidst swirling fluids and temperatures, as well as bubbles in equilibrium, as on the Chapter headings of this book. (See Section 16.8.)

In 2000, Hutchings, Morgan, Ritoré, and Ros announced a proof of the Double Bubble Conjecture, which says that the familiar double soap bubble provides the least-area way to enclose and separate two given volumes of air. The planar case was proved by my 1990 Williams College NSF "SMALL" undergraduate research Geometry Group [Foisy *et al.*]. The case of equal volumes in $\mathbf{R}^3$ was proved by Hass, Hutchings, and Schlafly with the help of computers in 1995. The general $\mathbf{R}^3$ proof was generalized to $\mathbf{R}^n$ by Reichardt. There are partial results in spheres, tori, and Gauss space, an important example of a manifold with density (see Chapters 18 and 19).

This little book provides the newcomer or graduate student with an illustrated introduction to geometric measure theory: the basic ideas, terminology, and results. It developed from my one-semester course at MIT for graduate students with a semester of graduate real analysis behind them. I have included a few fundamental arguments and a superficial discussion of the regularity theory, but my goal is merely to introduce the subject and make the standard text, Geometric Measure Theory by H. Federer, more accessible.

Other good references include L. Simon's Lectures on Geometric Measure Theory, E. Guisti's Minimal Surfaces and Functions of Bounded Variation, R. Hardt and Simon's Seminar on Geometric Measure Theory, Simon's Survey Lectures on Minimal Submanifolds, J. C. C. Nitsche's Lectures on

Minimal Surfaces (now available in English), R. Osserman's updated Survey of Minimal Surfaces, H. B. Lawson's Lectures on Minimal Submanifolds, A. T. Fomenko's books on The Plateau Problem, and S. Krantz and H. Parks's Geometric Integration Theory. S. Hildebrandt and A. Tromba offer a beautiful popular gift book for your friends, reviewed by Morgan [14, 15]. J. Brothers and also Sullivan and Morgan assembled lists of open problems. There is an excellent Questions and Answers about Area Minimizing Surfaces and Geometric Measure Theory by F. Almgren [4], who also wrote a review [5] of the first edition of this book. The easiest starting place may be the Monthly article "What is a Surface?" [Morgan 24].

It was from Fred Almgren, whose geometric perspective this book attempts to capture and share, that I first learned geometric measure theory. I thank many students for their interest and help, especially Benny Cheng, Gary Lawlor, Robert McIntosh, Mohamed Messaoudene, Marty Ross, Stephen Ai, David Ariyibi, John Bihn, John Herrera, Nam Nguyen, and Gabriel Ngwe. I also thank typists Lisa Court, Louis Kevitt, and Marissa Barschdorf. Jim Bredt first illustrated an article of mine as a member of the staff of Link, a one-time MIT student newspaper. I feel very fortunate to have him with me again on this book. I am grateful for help from many friends, notably Tim Murdoch, Yoshi Giga and his students, who prepared the Japanese translation, and especially John M. Sullivan. I would like to thank my new editor, Graham Nisbet, production manager Poulouse Joseph, and my original editor and friend Klaus Peters. A final thank you goes to all who contributed to this book at MIT, Rice, Stanford, and Williams. Some support was provided by National Science Foundation grants, by my Cecil and Ida Green Career Development Chair at MIT, and by my Dennis Meenan and Webster Atwell chairs at Williams.

This fifth edition includes updated material and references and a new Chapter 20 on the recently proved Log-Convex Density Theorem, one of many recent advances on manifolds with density and metric measure spaces.

Bibliographic references are simply by author's name, sometimes with an identifying numeral or section reference in brackets. Following a useful practice of Nitsche [2], the bibliography includes cross-references to each citation.

Frank Morgan
Williamstown, MA
fmorgan@williams.edu
http://math.williams.edu/morgan

PART I: BASIC THEORY

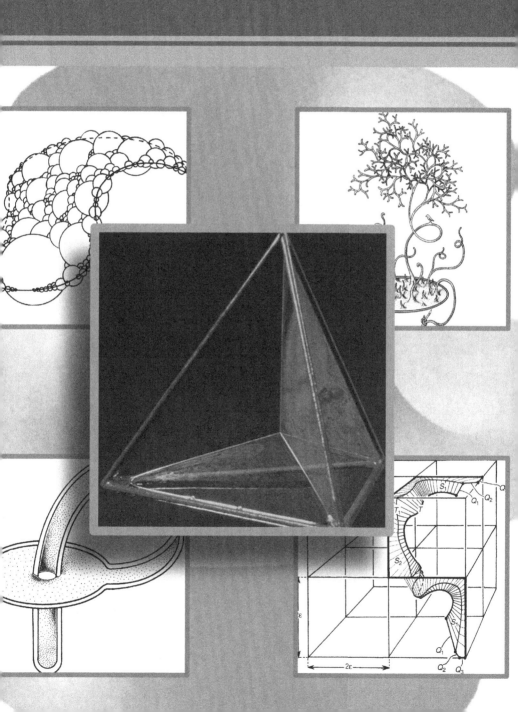

Geometric Measure Theory

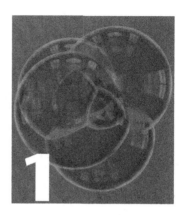

Geometric measure theory could be described as differential geometry, generalized through measure theory to deal with maps and surfaces that are not necessarily smooth, and applied to the calculus of variations. It dates from the 1960 foundational paper of Herbert Federer and Wendell Fleming on "Normal and Integral Currents," recognized by the 1986 AMS Steele Prize for a paper of fundamental or lasting importance, and earlier and contemporaneous work of L. C. Young [1, 2], E. De Giorgi [1, 3, 4], and E. R. Reifenberg [1–3] (see Figure 1.0.1). This chapter provides a rough outline of the purpose and basic concepts of geometric measure theory. Later chapters take up these topics more carefully.

1.1 Archetypical Problem Given a boundary in $\mathbf{R}^n$, find the surface of least area with that boundary. See Figure 1.1.1. Progress on this problem depends crucially on first finding a good space of surfaces to work in.

1.2 Surfaces as Mappings Classically, one considered only two-dimensional surfaces, defined as mappings of the disc. See Figure 1.2.1. Excellent references include J. C. C. Nitsche's *Lectures on Minimal Surfaces* [2], now available in English, R. Osserman's updated *Survey of Minimal Surfaces*, and H. B. Lawson's *Lectures on Minimal Submanifolds*. It was not until about 1930 that J. Douglas and T. Rado surmounted substantial inherent difficulties to prove that every smooth Jordan curve bounds a disc of least mapping area. Almost no progress was made for higher-dimensional surfaces (until, in a surprising turnaround, B. White [1] showed that for higher-dimensional surfaces the geometric measure theory solution actually solves the mapping problem too). The work of Douglas and Rado was generalized to certain smooth manifolds by C. B. Morrey and to proper metric spaces by A. Lytchak and S. Wenger.

Geometric Measure Theory. http://dx.doi.org/10.1016/B978-0-12-804489-6.00001-0

Figure 1.0.1. Wendell Fleming, Fred Almgren, and Ennio De Giorgi, three of the founders of geometric measure theory, at the Scuola Normale Superiore, Pisa, summer, 1965; and Fleming today. Photographs courtesy of Fleming.

Figure 1.1.1. The surface of least area bounded by two given Jordan curves.

Figure 1.2.1. Surface realized as a mapping, f, of the disc.

Along with its successes and advantages, the definition of a surface as a mapping has certain drawbacks (see Morgan [24]):

1. There is an inevitable *a priori* restriction on the types of singularities that can occur;
2. There is an *a priori* restriction on the topological complexity; and
3. The natural topology lacks compactness properties.

The importance of compactness properties appears in the direct method described in the next section.

1.3 The Direct Method

The direct method for finding a surface of least area with a given boundary has three steps.

1. Take a sequence of surfaces with areas decreasing to the infimum.
2. Extract a convergent subsequence.
3. Show that the limit surface is the desired surface of least area.

Figures 1.3.1–1.3.4 show how this method breaks down for lack of compactness in the space of surfaces as mappings, even when the given boundary is the unit

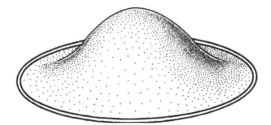

Figure 1.3.1. A surface with area $\pi + 1$.

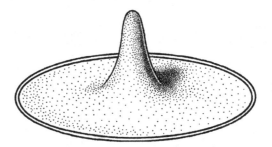

Figure 1.3.2. A surface with area $\pi + \frac{1}{4}$.

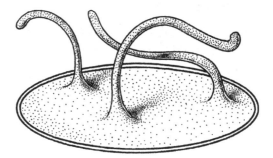

Figure 1.3.3. A surface with area $\pi + \frac{1}{16}$.

circle. By sending out thin tentacles toward every rational point, the sequence could include all of $\mathbf{R}^3$ in its closure!

1.4   Rectifiable Currents   An alternative to surfaces as mappings is provided by *rectifiable currents*, the m-dimensional, oriented surfaces of geometric measure theory. The relevant functions $f : \mathbf{R}^m \to \mathbf{R}^n$ need not be smooth but merely *Lipschitz*; that is,

$$|f(x) - f(y)| \leq C|x - y|,$$

for some "Lipschitz constant" C.

Fortunately, there is a good m-dimensional measure on $\mathbf{R}^n$, called *Hausdorff measure*, $\mathcal{H}^m$. Hausdorff measure agrees with the classical mapping area of an embedded manifold, but it is defined for all subsets of $\mathbf{R}^n$.

A Borel subset B of $\mathbf{R}^n$ is called $(\mathcal{H}^m, m)$ *rectifiable* if B is a countable union of Lipschitz images of bounded subsets of $\mathbf{R}^m$, with $\mathcal{H}^m(B) < \infty$. (As usual,

Figure 1.3.4. A surface with area $\pi + \frac{1}{64}$.

we will ignore sets of $\mathcal{H}^m$ measure 0.) That definition sounds rather general, and it includes just about any "m-dimensional surface" I can imagine. Nevertheless, these sets will support a kind of differential geometry: for example, it turns out that a rectifiable set B has a canonical tangent plane at almost every point.

Finally, a *rectifiable current* is an oriented rectifiable set with integer multiplicities, finite area, and compact support. By general measure theory, one can integrate a smooth differential form φ over such an oriented rectifiable set S, and hence view S as a *current*; that is, a linear functional on differential forms,

$$\varphi \mapsto \int_S \varphi.$$

This perspective yields a new natural topology on the space of surfaces, dual to an appropriate topology on differential forms. This topology has useful compactness properties, given by the fundamental compactness theorem in Section 1.5. Viewing rectifiable sets as currents also provides a boundary operator ∂ from m-dimensional rectifiable currents to $(m - 1)$-dimensional currents, defined by

$$(\partial S)(\varphi) = S(d\varphi),$$

where $d\varphi$ is the exterior derivative of φ. By Stokes's theorem, this definition coincides with the usual notion of boundary for smooth, compact, oriented manifolds with boundary. In general, the current ∂S is not rectifiable, even if S is rectifiable.

1.5   The Compactness Theorem   *Let c be a positive constant. Then the set of all m-dimensional rectifiable currents T in a fixed large closed ball in $\mathbf{R}^n$, such that the boundary ∂T is also rectifiable and such that the area of both T and ∂T is bounded by c, is compact in an appropriate weak topology.*

1.6   Advantages of Rectifiable Currents   Notice that rectifiable currents have none of the three drawbacks mentioned in Section 1.2. There is certainly no restriction on singularities or topological complexity. Moreover, the compactness theorem provides the ideal compactness properties. In fact, the direct method described in Section 1.3 succeeds in the context of rectifiable currents. In the figures of Section 1.3, the amount of area in the tentacles goes to 0. Therefore, they disappear in the limit in the new topology. What remains is the disc, the desired solution.

All of these results hold in all dimensions and codimensions.

1.7   The Regularity of Area-Minimizing Rectifiable Currents
One serious suspicion hangs over this new space of surfaces: The solutions they provide to the problem of least area, the so-called area-minimizing rectifiable currents, may be generalized objects without any geometric significance. The following interior regularity results allay such concerns. (We give more precise statements in Chapter 8.)

1. A two-dimensional area-minimizing rectifiable current in $\mathbf{R}^3$ is a smooth embedded manifold.
2. For $m \leq 6$, an m-dimensional area-minimizing rectifiable current in $\mathbf{R}^{m+1}$ is a smooth embedded manifold.

Thus, in low dimensions the area-minimizing hypersurfaces provided by geometric measure theory actually turn out to be smooth embedded manifolds. However, in higher dimensions, singularities occur, for geometric and not merely technical reasons (see Section 10.7). Despite marked progress, understanding such singularities remains a tremendous challenge.

1.8 More General Ambient Spaces

Basic geometric measure theory extends from $\mathbf{R}^n$ to Riemannian manifolds via C^1 embeddings in $\mathbf{R}^n$ or Lipschitz charts. Ambrosio and Kirchheim among others have been developing an intrinsic approach to geometric measure theory in certain metric spaces.

Measures

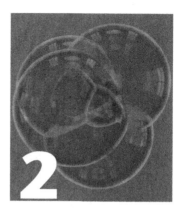

This chapter lays the measure-theoretic foundation, including the definition of Hausdorff measure and covering theory. The general reference is Federer [1, Chapter II].

2.1 Definitions For us, a *measure* μ on $\mathbf{R}^n$ will be what is sometimes called an outer measure: a nonnegative function μ on *all* subsets of $\mathbf{R}^n$ (with the value $+\infty$ allowed, of course), which is *countably subadditive*; that is, if A is contained in a countable union, $\cup A_i$, then

$$\mu(A) \le \sum \mu(A_i).$$

A set $A \subset \mathbf{R}^n$ is called *measurable* if, for all $E \subset \mathbf{R}^n$, $\mu(E \cap A) + \mu(E \cap A^C) = \mu(E)$. The class of measurable sets is a σ-algebra—that is, closed under complementation, countable union, and countable intersection. If A is a countable disjoint union of measurable sets A_i, then $\mu(A) = \sum \mu(A_i)$.

The smallest σ-algebra containing all open sets is the collection of *Borel* sets. A measure μ is called *Borel regular* if Borel sets are measurable and every subset of $\mathbf{R}^n$ is contained in a Borel set of the same measure.

Suppose that μ is Borel regular, A is measurable, and $\varepsilon > 0$. If $\mu(A) < \infty$, then A contains a closed subset C with $\mu(A - C) < \varepsilon$. If A can be covered by countably many open sets of finite measure, then A is contained in an open set W with $\mu(W - A) < \varepsilon$ [Federer, 2.2.3].

All Borel sets are measurable if and only if Caratheodory's criterion holds:

Whenever A_1, A_2 are sets a positive distance apart, then

(1) $$\mu(A_1 \cup A_2) = \mu(A_1) + \mu(A_2).$$

Geometric Measure Theory. http://dx.doi.org/10.1016/B978-0-12-804489-6.00002-2

2.2 Lebesgue Measure There is a unique Borel regular, translation invariant measure on $\mathbf{R}^n$ such that the measure of the unit cube $[0,1]^n$ is 1. This measure is called *Lebesgue measure, $\mathscr{L}^n$*.

2.3 Hausdorff Measure [Federer, 2.10], [Falconer] Unfortunately, for general "m-dimensional" subsets of $\mathbf{R}^n$ (for $m < n$), it is more difficult to assign an m-dimensional measure. The m-dimensional area of a C^1 map f from a domain $D \subset \mathbf{R}^m$ into $\mathbf{R}^n$ is classically defined as the integral of the Jacobian $J_m f$ over D. [Computationally, at each point $x \in D$, $(J_m f)^2$ equals the sum of the squares of the determinants of the $m \times m$ submatrices of $Df(x)$ or, equivalently, the determinant of $(Df(x))^t Df(x)$.] The area of an m-dimensional submanifold M of $\mathbf{R}^n$ is then defined by calculating it on parameterized portions of M and proving that the area is independent of choice of parameterization.

In 1918, F. Hausdorff introduced an m-dimensional measure in $\mathbf{R}^n$ that gives the same area for submanifolds but is defined on all subsets of $\mathbf{R}^n$. When $m = n$, it turns out to be equal to Lebesgue measure.

Definitions For any subset S of $\mathbf{R}^n$, define the *diameter* of S

$$\text{diam}(S) = \sup\{|x - y|: x, y \in S\}.$$

Let α_m denote the Lebesgue measure of the closed unit ball $\mathbf{B}^m(0, 1) \subset \mathbf{R}^m$. For $A \subset \mathbf{R}^n$, we define the m-dimensional Hausdorff measure $\mathscr{H}^m(A)$ by the following process. For small δ, cover A efficiently by countably many sets S_j with $\text{diam}(S_j) \leq \delta$, add up all the $\alpha_m(\text{diam}(S_j)/2)^m$, and take the limit as $\delta \to 0$:

$$\mathscr{H}^m(A) = \lim_{\delta \to 0} \inf_{\substack{A \subset \bigcup S_j \\ diam(S_j) \leq \delta}} \sum \alpha_m \left(\frac{\text{diam}(S_j)}{2} \right)^m.$$

The infimum is taken over all countable coverings $\{S_j\}$ of A whose members have diameter at most δ. As δ decreases, the more restricted infimum cannot decrease, and hence the limit exists, with $0 \leq \mathscr{H}^m(A) \leq \infty$. In Figure 2.3.1, the two-dimensional area is approximated by $\sum \pi r^2$. The spiral of Figure 2.3.2 illustrates one reason for taking the limit as $\delta \to 0$, since otherwise a spiral of great length could be covered by a single ball of radius 1.

Countable subadditivity follows immediately from the definition. The measurability of Borel sets follows easily from Caratheodory's criterion 2.1(1).

To see that each $A \subset \mathbf{R}^n$ is contained in a Borel set B of the same measure, note first that each S_j occurring in the definition of $\mathscr{H}^m(A)$ may be replaced by its closure so that $\cup S_j$ is Borel. If $\{S_j^{(k)}\}$ is a countable sequence of coverings

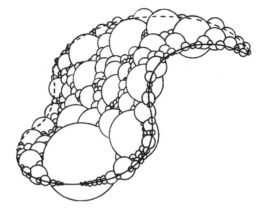

Figure 2.3.1. The Hausdorff measure (area) of a piece of surface A is approximated by the cross sections of little balls that cover it.

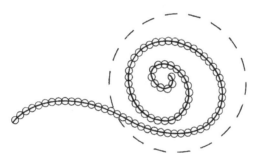

Figure 2.3.2. One must cover by *small* sets to compute length accurately. Here, the length of the spiral is well estimated by the sum of the diameters of the tiny balls, but it is grossly underestimated by the diameter of the huge ball.

defining $\mathscr{H}^m(A)$, then $B = \cap_k \cup_j S_j^{(k)}$ gives the desired Borel set. Therefore, $\mathscr{H}^m$ is Borel regular. Later, it will be proved that $\mathscr{H}^m$ gives the "correct" area for C^1 submanifolds of $\mathbf{R}^n$.

The definition of Hausdorff measure extends to any nonnegative real dimension. [The definition of α_m is extended by the Γ function: $\alpha_m = \pi^{m/2} / \Gamma(m/2 + 1)$]. Notice that $\mathscr{H}^0$ is counting measure: $\mathscr{H}^0(A)$ is the number of elements of A.

The *Hausdorff dimension* of a nonempty set A is defined as

$$\inf(m \geq 0: \mathscr{H}^m(A) < \infty) = \inf\{m: \mathscr{H}^m(A) = 0\}$$
$$= \sup\{m: \mathscr{H}^m(A) > 0\}$$
$$= \sup\{m: \mathscr{H}^m(A) = \infty\}.$$

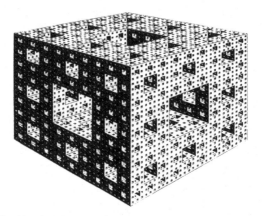

Figure 2.3.3. The Sierpinski sponge is an example of a fractional dimensional set. Its Hausdorff dimension is approximately 2.7. (From *Studies in Geometry* by Leonard M. Blumenthal and Karl Menger. Copyright 1979 by W. H. Freeman and Company. Reprinted with permission.)

The equivalence of these conditions follows from the fact that if $m < k$ and $\mathscr{H}^m(A) < \infty$, then $\mathscr{H}^k(A) = 0$ (Exercise 2.4). The Cantor set of Exercise 2.6 turns out to have Hausdorff dimension $\ln 2 / \ln 3$. Figure 2.3.3 pictures a Cantor-like set in $\mathbf{R}^3$, called the Sierpinski sponge, which has Hausdorff dimension of approximately 2.7.

These Cantor-like sets are self-similar in the sense that certain homothetic expansions of such a set are locally identical to the original set. Self-similarity appears in the coastline of Great Britain and in the mass in the universe. B. Mandelbrot has modeled many natural phenomena by random fractional dimensional sets and processes, called *fractals*. His books, *Fractals* and *The Fractal Geometry of Nature,* contain beautiful, computer-generated pictures of hypothetical clouds, landscapes, and other phenomena.

2.4 Integral-Geometric Measure

In 1932, J. Favard defined another m-dimensional measure on $\mathbf{R}^n (m = 0, 1, \ldots, n)$, now called integral-geometric measure, $\mathscr{I}^m$. It turns out that $\mathscr{I}^m$ agrees with $\mathscr{H}^m$ on all smooth m-dimensional submanifolds and other nice sets, but it disagrees and often is zero on Cantor-like sets.

Roughly, to define $\mathscr{I}^m(A)$, project A onto an m-dimensional subspace of $\mathbf{R}^n$, take the Lebesgue measure (counting multiplicities), and average over all such projections.

More precisely, let $\mathbf{O}^*(n, m)$ denote the set of orthogonal projections p of $\mathbf{R}^n$ onto m-dimensional subspaces. For general reasons there is a unique measure on $\mathbf{O}^*(n, m)$ invariant under Euclidean motions on $\mathbf{R}^n$, normalized to have total measure 1. For example, the set $O^*(2, 1)$ of orthogonal projections onto lines through 0 in the plane may be parameterized by $0 \leq \theta < \pi$, and the unique measure is $(1/\pi)d\theta$. For $y \in$ image $p \cong \mathbf{R}^m$, let the "multiplicity function," $N(p|A, y)$, denote the number of points in $A \cap p^{-1}(y)$. Define a normalizing constant,

$$\beta(n, m) = \Gamma\left(\frac{m+1}{2}\right) \Gamma\left(\frac{n-m+1}{2}\right) \Gamma\left(\frac{n+1}{2}\right)^{-1} \pi^{-1/2}.$$

Now define the integral-geometric measure of any Borel set B by

$$\mathscr{I}^m(B) = \frac{1}{\beta(n, m)} \int_{p \in O^*(n, m)} \int_{y \in \text{image } p \cong \mathbf{R}^m} N(p|B, y) d\,\mathscr{L}^m y \, dp.$$

One checks that the function $N(p|B, y)$ is indeed measurable and that $\mathscr{I}^m$ is countably subadditive. Finally, extend $\mathscr{I}^m$ to a Borel regular measure by defining for any set $A \subset \mathbf{R}^n$,

$$\mathscr{I}^m(A) = \inf\{\mathscr{I}^m(B): A \subset B, \ B \text{ Borel}\}.$$

2.5   Densities [Federer, 2.9.12, 2.10.19]   Let A be a subset of $\mathbf{R}^n$. For $1 \leq m \leq n$, $a \in \mathbf{R}^n$, we define the m-dimensional *density* $\Theta^m(A, a)$ of A at a by the formula

$$\Theta^m(A, a) = \lim_{r \to 0} \frac{\mathscr{H}^m(A \cap \mathbf{B}^n(a, r))}{\alpha_m r^m},$$

where α_m is the measure of the closed unit ball $\mathbf{B}^m(0, 1)$ in $\mathbf{R}^m$. (This limit may or may not exist.) For example, the cone

$$C = \{x^2 + y^2 = z^2\}$$

of Figure 2.5.1 has two-dimensional density

$$\Theta^2(C, a) = \begin{cases} 1 & \text{for} \quad a \in C - \{0\}, \\ 0 & \text{for} \quad a \notin C, \\ \sqrt{2} & \text{for} \quad a = 0. \end{cases}$$

Similarly, for μ a measure on $\mathbf{R}^n$, $1 \leq m \leq n$, $a \in \mathbf{R}^n$, define the m-dimensional density $\Theta^m(\mu, a)$ of μ at a by

$$\Theta^m(\mu, a) = \lim_{r \to 0} \frac{\mu(\mathbf{B}^n(a, r))}{\alpha_m r^m}.$$

Figure 2.5.1. The cone $\{x^2 + y^2 = z^2\}$ has density 1 everywhere except at the vertex, where it has density $\sqrt{2}$.

Note that for any subset A of $\mathbf{R}^n$, $\Theta^m(A, a) = \Theta^m(\mathscr{H}^m \llcorner A, a)$, where $\mathscr{H}^m \llcorner A$ is the measure defined by

$$(\mathscr{H}^m \llcorner A)(E) \equiv \mathscr{H}^m(A \cap E).$$

Hence, density of measures actually generalizes the notion of density of sets.

2.6 Approximate Limits [Federer, 2.9.12]

Let $A \subset \mathbf{R}^m$. A function $f: A \to \mathbf{R}^n$ has *approximate limit* y at a if for every $\varepsilon > 0$, $\mathbf{R}^m - \{x \in A: |f(x) - y| < \varepsilon\}$ has m-dimensional density 0 at a. We write $y = ap\lim_{x \to a} f(x)$. Note that in particular A must have density 1 at a.

Proposition *A function* $f: A \to \mathbf{R}^n$ *has an approximate limit* y *at* a *if and only if there is a set* $B \subset A$ *such that* B^C *has m-dimensional density* 0 *at* a *and* $f|B$ *has the limit* y *at* a.

Remark In general, the word *approximate* means "except for a set of density 0."

Proof The condition is clearly sufficient. To prove necessity, assume that f has an approximate limit y at a. For convenience, we assume $y = 0$. Then for any positive integer i,

$$A_i \equiv \mathbf{R}^m - \{x \in A: |f(x)| < 1/i\}$$

has density 0 at a. Choose $r_1 > r_2 > \ldots$ such that

$$\frac{\mathscr{H}^m(A_i \cap \mathbf{B}^n(a, r))}{\alpha_m r^m} \leq 2^{-i}$$

whenever $0 < r \leq r_i$. Notice that $A_1 \subset A_2 \subset \ldots$. Let $B^C = \cup(A_i \cap \mathbf{B}(a, r_i))$.

Clearly, $f|B$ has the limit y at a. To show that B^C has density 0 at a, let $r_i > s > r_{i+1}$. Then

$$\mathcal{H}^m(B^C \cap \mathbf{B}(a, s)) \leq \mathcal{H}^m(A_i \cap \mathbf{B}(a, s)) + \mathcal{H}^m(A_{i+1} \cap \mathbf{B}(a, r_{i+1}))$$
$$+ \mathcal{H}^m(A_{i+2} \cap \mathbf{B}(a, r_{i+2})) + \cdots$$
$$\leq \alpha_m(s^m \cdot 2^{-i} + r_{i+1}^m \cdot 2^{-(i+1)} + r_{i+2}^m \cdot 2^{-(i+2)} + \cdots)$$
$$\leq \alpha_m \cdot s^m \cdot 2^{-(i-1)}.$$

Therefore, B^C has density 0 at a, as desired.   ∎

Definitions Let $a \in A \subset \mathbf{R}^m$. A function $f : A \to \mathbf{R}^n$ is *approximately continuous* at a if $f(a) = ap \lim_{x \to a} f(x)$. The point a is a *Lebesgue point* of f if $\Theta^m(A^C, a) = 0$ and

$$\frac{1}{\alpha_m r^m} \int_{A \cap \mathbf{B}(a, r)} |f(x) - f(a)| \, d\mathscr{L}^m x \xrightarrow[r \to 0]{} 0.$$

The function f is *approximately differentiable* at a if there is a linear function $L: \mathbf{R}^m \to \mathbf{R}^n$ such that

$$ap \lim_{x \to a} \frac{|f(x) - f(a) - L(x - a)|}{|x - a|} = 0.$$

We write $L = ap Df(a)$.

The following covering theorem of Besicovitch proves more powerful in practice than more familiar ones, such as that of Vitali. It applies to any finite Borel measure φ.

2.7 Besicovitch Covering Theorem [Federer, 2.8.15; Besicovitch] *Suppose φ is a Borel regular measure on $\mathbf{R}^n$, $A \subset \mathbf{R}^n$, $\varphi(A) < \infty$, F is a collection of nontrivial closed balls, and $\inf\{r: \mathbf{B}(a, r) \in F\} = 0$ for all $a \in A$. Then there is a (countable) disjoint subcollection of F that covers φ almost all of A.*

Partial Proof We may assume that all balls in F have radius at most 1.

PART 1 *There is a constant $\zeta(n)$ such that, given a closed ball, B, of radius r and a collection, C, of closed balls of a radius of at least r which intersect B and which do not contain each other's centers, then the cardinality of C is at most $\zeta(n)$.* This statement is geometrically obvious, and we omit the proof. E. R. Reifenberg [4] proved that for $n = 2$, the sharp bound is 18 (see Figure 2.7.1).

Figure 2.7.1. At most, 18 larger discs can intersect the unit disc in $\mathbf{R}^2$ without containing each other's centers. Figure courtesy of J. M. Sullivan [2].

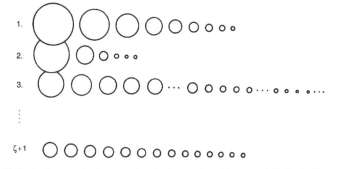

Figure 2.7.2. In the proof of the Besicovitch covering Theorem 2.7, the balls covering A are arranged by size in rows or discarded. Intersecting balls must go in different rows. For the case of $\mathbf{R}^2$, this requires at most $\zeta(2) + 1 = 19$ rows. Then some row must provide a disjoint cover of at least $1/19$ the total measure.

PART 2 $\zeta + 1$ *subcollections of disjoint balls cover A.* To prove this statement, we will arrange the balls of F in rows of disjoint balls, starting with the largest and proceeding in order of size. (Of course, there may not always be a "largest ball," and actually one chooses a nearly largest ball. This technical point propagates minor corrections throughout the proof, which we will ignore.)

Place the largest ball B_1 in the first row. (See Figure 2.7.2.) Throw away all balls whose centers are covered by B_1.

Take the next largest ball, B_2. If B_2 is disjoint from B_1, place B_2 in the first row. If not, place B_2 in the second row. Throw away all balls whose centers are covered by B_2.

At the nth step, place B_n in the earliest row that keeps all balls in each row disjoint. Throw away all balls whose centers are covered by B_n.

Proceed by transfinite induction. The whole list certainly covers A, since we throw away only balls whose centers are already covered. Each row consists of disjoint balls, by construction. Hence, it suffices to show that there are at most $\zeta + 1$ nonempty rows. Assume some ball, B, gets put in the $\zeta + 2$ row. Then there are balls $D_1, \ldots, D_{\zeta+1}$, at least as large as B already in the first $\zeta + 1$ rows and not disjoint from B. No D_j can contain another's center, or the smaller would have been thrown away when the larger was put in. This contradiction of Part 1 completes the proof of Part 2.

PART 3 *Completion of proof.* By Part 2, some disjoint subcollection covers $1/(\zeta + 1)$ the φ measure of A. Hence, some finite disjoint subcollection covers a closed subset $A_1 \subset A$ with

$$\frac{\varphi(A_1)}{\varphi(A)} \geq \frac{1}{\zeta + 2}, \text{ i.e., } 1 - \frac{\varphi(A_1)}{\varphi(A)} \leq \delta < 1.$$

Repeat the whole process on $A - A_1$ with the balls contained in $A - A_1$ to obtain a finite disjoint subcollection covering $A_2 \subset A$ with

$$1 - \frac{\varphi(A_2)}{\varphi(A)} \leq \delta^2.$$

Countably many such repetitions finally yield a countable disjoint subcollection covering φ almost all of A. ∎

We now give three corollaries as examples of the usefulness of Besicovitch's covering theorem.

2.8   Corollary $\mathcal{H}^n = \mathcal{L}^n$ on $\mathbf{R}^n$.

Proof We will need the so-called isodiametric inequality or Bieberbach inequality, which says that among all sets of fixed diameter, the ball has the largest volume. In other words, for any set S in $\mathbf{R}^n$,

$$\mathcal{L}^n(S) \leq \alpha_n \left(\frac{\text{diam } S}{2} \right)^n.$$

It follows immediately that $\mathcal{H}^n \geq \mathcal{L}^n$.

The original proof of Bieberbach for the planar case uses a wonderful symmetrization argument (and unnecessary convexification). This argument works in general dimension. Later, it will be the key to proving that area-minimizing double bubbles are rotationally symmetric (Thm. 14.3). We may assume that S is symmetric with respect to each coordinate axis since replacing each intersection of S with a line parallel to the axis by a symmetric interval of the same one-dimensional measure does not change the Lebesgue measure and can only decrease the diameter. (Such symmetrization is called Steiner symmetrization or Schwarz symmetrization, which coincide in the plane; see §13.2.) But now S is symmetric with respect to the origin and hence is contained in the ball B of the same diameter. Therefore,

$$\mathcal{L}^n(S) \leq \mathcal{L}^n(B) = \alpha_n \left(\frac{\text{diam } S}{2} \right)^n,$$

as desired. Notice that the symmetrization step is necessary because an equilateral triangle, for example, is not contained in a ball of the same diameter.

On the proof of the isodiametric inequality, see also Blaschke [§25.IV, pp. 122–123], Burago and Zalgaller [Thm. 11.2.1], and Federer [1, Cor. 2.10.33].

To prove that $\mathcal{H}^n \leq \mathcal{L}^n$, we will use the Besicovitch covering theorem. First, we note that it suffices to prove that $\mathcal{H}^n(A) \leq \mathcal{L}^n(A)$ for A Borel and bounded, or hence for A equal to the open R-ball $\mathbf{U}^n(0, R) \subset \mathbf{R}^n$, or hence for $A = \mathbf{U}^n(0, 1)$. An easy computation shows that $\mathcal{H}^n(A) < \infty$. Given $\varepsilon > 0$, choose $\delta > 0$ such that

(1) $\quad \mathcal{H}^n(A) \leq \inf \left\{ \sum \alpha_n \cdot \left(\frac{\text{diam } S_i}{2} \right)^n : A \subset \cup S_i, \text{ diam } S_i \leq \delta \right\} + \varepsilon.$

Apply the covering theorem with

$$F = \{\text{closed balls contained in } A \text{ with diameter} \leq \delta\}$$

to obtain a disjoint covering G of $B \subset A$ with $\mathcal{H}^n(A - B) = 0$. Let G' be a covering by balls of diameter at most δ of $A - B$ with

$$\sum_{S \in G'} \alpha_n \left(\frac{\text{diam } S}{2} \right)^n \leq \varepsilon.$$

Then $G \cup G'$ covers A, and therefore

$$\mathcal{H}^n(A) \leq \sum_{S \in G \cup G'} \alpha_n \left(\frac{\text{diam } S}{2} \right)^n + \varepsilon$$

$$\leq \sum_{S \in G} \mathcal{L}^n(S) + \sum_{S \in G'} \alpha_n \left(\frac{\text{diam } S}{2} \right)^n + \varepsilon$$

$$\leq \mathcal{L}^n(A) + \varepsilon + \varepsilon.$$

The corollary is proved. The fussing with $A - B$ at the end was necessary because (1) does not apply to B. ∎

2.9   Corollary   *If $A \subset R^n$ is Lebesgue measurable, then the density $\Theta^n(A, x)$ equals the characteristic function $\chi_A(x)$ almost everywhere.*

Proof   It suffices to show that for every measurable set A, $\Theta(A, x) = 1$ at almost all points $x \in A$. [Considering A^C then implies $\Theta(A, x) = 0$ at almost all $x \notin A$.] Assume not. We may assume $0 < \mathscr{L}^n(A) < \infty$. We may further assume that for some $\delta < 1$

(1) $\displaystyle \Theta_*(A, a) = \varliminf \frac{\mathscr{L}^n(A \cap \mathbf{B}(a, r))}{\alpha_n r^n} < \delta$ for all $a \in A$,

by first choosing δ such that

$$\mathscr{L}^n \{a \in A : \Theta_*(A, a) < \delta\} > 0$$

and then replacing A by $\{a \in A : \Theta_*(A, a) < \delta\}$. Choose an open set $U \supset A$ such that

(2) $\mathscr{L}^n(A) > \delta \mathscr{L}^n(U)$.

Let F be the collection of all closed balls B centered in A and contained in U such that

$$\mathscr{L}^n(A \cap B) < \delta \mathscr{L}^n(B).$$

By (1), F contains arbitrarily small balls centered at each point of A. By the covering theorem, there is a countable disjoint subcollection G covering almost all of A. Therefore,

$$\mathscr{L}^n(A) < \delta \sum_{B \in G} \mathscr{L}^n B \leq \delta \mathscr{L}^n(U).$$

This contradiction of (2) proves the corollary. ∎

2.10   Corollary   *A measurable function $f: \mathbf{R}^n \to \mathbf{R}$ is approximately continuous almost everywhere.*

Corollary 2.10 follows rather easily from Corollary 2.9. Exercise 2.9 gives some hints on the proof.

Exercises

2.0   Prove that a positive scaling rA of a set A in $\mathbf{R}^n$ satisfies $\mathscr{H}^m(rA) = r^m \mathscr{H}^m(A)$.

2.1 Let I be the line segment in $\mathbf{R}^2$ from $(0, 0)$ to $(1, 0)$. Compute $\mathscr{I}^1(I)$ directly. $[\beta(2, 1) = 2/\pi.]$

2.2 Let I be the unit interval $[0, 1]$ in $\mathbf{R}^1$. Prove that $\mathscr{H}^1(I) = 1$.

2.3 Prove that $\mathscr{H}^n(\mathbf{B}^n(0, 1)) < \infty$, just using the definition of Hausdorff measure.

2.4 Let A be a nonempty subset of $\mathbf{R}^n$. First, prove that if $0 \le m < k$ and $\mathscr{H}^m(A) < \infty$, then $\mathscr{H}^k(A) = 0$. Second, deduce that the four definitions of the Hausdorff dimension of A are equivalent.

2.5 Define a set $A \subset \mathbf{R}^2$ as in the following figure by starting with an equilateral triangle and removing triangles as follows. Let A_0 be a closed equilateral triangular region of side 1. Let A_1 be the three equilateral triangular regions of side $1/3$ in the corners of A_0. In general, let A_{j+1} be the triangular regions, a third the size, in the corners of the triangles of A_j. Let $A = \cap A_j$. Prove that $\mathscr{H}^1(A) = 1$.

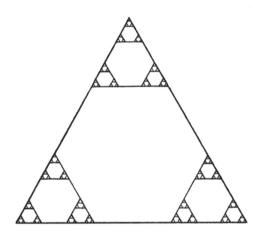

2.6 To define the usual Cantor set $C \subset \mathbf{R}^1$, let $C_1 = [0, 1]$; construct C_{j+1} by removing the open middle third of each interval of C_j and put

$$C = \cap\{C_j : j \in \mathbf{Z}^+\}.$$

Let $m = \ln 2 / \ln 3$.

a. Prove that $\mathscr{H}^m(C) \le \alpha_m / 2^m$ and, hence, $\dim C \le m$.

b. Try to prove that $\mathscr{H}^m(C) = \alpha_m / 2^m$ or at least that $\mathscr{H}^m(C) > 0$ and hence that the Hausdorff dimension of C is m.

2.7 Give a function $f : \mathbf{R}^2 \to \mathbf{R}$ which is approximately continuous at 0, but for which 0 is not a Lebesgue point.

2.8 Prove that if $f: \mathbf{R}^m \to \mathbf{R}$ has $\mathbf{0}$ as a Lebesgue point, then f is approximately continuous at $\mathbf{0}$.

2.9 Deduce Corollary 2.10 from Corollary 2.9.

Hint: Let $\{q_i\}$ be a countable dense subset of $\mathbf{R}$, $A_i = \{x: f(x) > q_i\}$, and $E_i = \{x: \Theta(A_i, x) = \chi_{A_i}\}$, and show that f is approximately continuous at each point in $\cap E_i$.

Lipschitz Functions and Rectifiable Sets

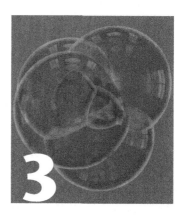

This chapter introduces the m-dimensional surfaces of geometric measure theory, called rectifiable sets. These sets have folds, corners, and more general singularities. The relevant functions are not smooth functions as in differential geometry, but *Lipschitz* functions. See also the survey, "What Is a Surface?" [Morgan 24].

3.1 Lipschitz Functions

A function $f : \mathbf{R}^m \to \mathbf{R}^n$ is *Lipschitz* if there is a constant C such that

$$|f(x) - f(y)| \leq C|x - y|.$$

The least such constant is called the *Lipschitz constant* and is denoted by Lip f. Figure 3.1.1 gives the graphs of two typical Lipschitz functions. Theorems 3.2 and 3.3 show that a Lipschitz function comes very close to being differentiable.

3.2 Rademacher's Theorem [Federer, 3.1.6]

A Lipschitz function $f : \mathbf{R}^m \to \mathbf{R}^n$ is differentiable almost everywhere.

The *proof* has five steps:

(1) A monotonic function $f : \mathbf{R} \to \mathbf{R}$ is differentiable almost everywhere.
(2) Every function $f : \mathbf{R} \to \mathbf{R}$ which is locally of bounded variation (and hence every Lipschitz function) is differentiable almost everywhere.
(3) A Lipschitz function $f : \mathbf{R}^m \to \mathbf{R}^n$ has partial derivatives almost everywhere.
(4) A Lipschitz function $f : \mathbf{R}^m \to \mathbf{R}^n$ is approximately differentiable almost everywhere.
(5) A Lipschitz function $f : \mathbf{R}^m \to \mathbf{R}^n$ is differentiable almost everywhere.

Geometric Measure Theory. http://dx.doi.org/10.1016/B978-0-12-804489-6.00003-4

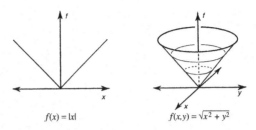

$f(x) = |x|$ $f(x, y) = \sqrt{x^2 + y^2}$

Figure 3.1.1. Examples of Lipschitz functions.

Step (1) is a standard result of real analysis, proved by differentiation of measures. Step (2) follows by decomposing a function of bounded variation as the difference of two monotonic functions. Step (3) follows immediately from Step (2) (modulo checking measurability). The deduction of (4) from (3) is technical, but not surprising, because the existence of continuous partial derivatives implies differentiability and a measurable function is approximately continuous almost everywhere. If (3) holds everywhere, it does not follow that (4) holds everywhere.

The final conclusion (5) rests on the interesting fact that *if a Lipschitz function is approximately differentiable at a, it is differentiable at a.* We conclude this discussion with a proof of that fact.

Suppose that the Lipschitz function $f : \mathbf{R}^m \to \mathbf{R}^n$ is approximately differentiable at a but not differentiable at a. We may assume $a = 0$, $f(0) = 0$, and $a \, p \, Df(0) = 0$.

For some $0 < \varepsilon < 1$, there is a sequence of points $a_i \to 0$ such that

$$|f(a_i)| \geq \varepsilon |a_i|.$$

Let $C = \max\{\text{Lip } f, 1\}$. Then, for x in the closed ball $\mathbf{B}(a_i, \varepsilon|a_i|/3C)$,

$$|f(x)| \geq \varepsilon |a_i| - \varepsilon |a_i|/3 \geq \varepsilon |x|/2.$$

Thus, for $x \in E = \cup_{i=1}^{\infty} \mathbf{B}(a_i, \varepsilon|a_i|/3C)$,

$$|f(x)| \geq \varepsilon |x|/2.$$

But E does not have density 0 at $\mathbf{0}$ because

$$\frac{\mathscr{L}^m \mathbf{B}(a_i, \varepsilon|a_i|/3C)}{\alpha_m(|a_i| + \varepsilon|a_i|/3C)^m} \geq \frac{(\varepsilon|a_i|/3C)^m}{(4|a_i|/3)^m} = \frac{\varepsilon^m}{4^m C^m} > 0.$$

This contradiction of the approximate differentiability of f at $\mathbf{0}$ completes the proof. ∎

3.3   Approximation of a Lipschitz Function by a C^1 Function

[Federer, 3.1.15]   *Suppose that $A \subset \mathbf{R}^m$ and that $f : A \to \mathbf{R}^n$ is Lipschitz. Given $\varepsilon > 0$, there is a C^1 function $g : \mathbf{R}^m \to \mathbf{R}^n$ such that $\mathscr{L}^m \{x \in A : f(x) \neq g(x)\} \leq \varepsilon$.*

Note that the approximation is in the strongest sense: The functions *coincide* except on a set of measure ε. The proof of 3.3 depends on Whitney's extension theorem, which gives the coherence conditions on prescribed values for a desired C^1 function.

3.4   Lemma (Whitney's Extension Theorem) [Federer, 3.1.14]

Let A be a closed set of points a in $\mathbf{R}^m$ at which the values and derivatives of a desired C^1 function are prescribed by linear polynomials $P_a : \mathbf{R}^m \to \mathbf{R}$. For each compact subset C of A and $\delta > 0$, let $\rho(C, \delta)$ be the supremum of the numbers $|P_a(b) - P_b(b)|/|a - b|$, $\|DP_a(b) - DP_b(b)\|$, over all $a, b \in C$ with $0 < |a - b| \leq \delta$. If the prescribed data satisfy the coherence condition that $\lim_{\delta \to 0} \rho(C, \delta) = 0$ for each compact subset C of A, then there exists a C^1 function g satisfying

$$g(a) = P_a(a), \ Dg(a) = DP_a(a)$$

for all $a \in A$.

Remarks   A more general version of Whitney's extension theorem gives the analogous conditions to obtain a C^k function with values and derivatives prescribed by polynomials P_a of degree k. In the proof, the value $g(x)$ assigned at a point $x \notin A$ is a smoothly weighted average of the values prescribed at nearby points of A. The averaging uses a partition of unity subordinate to a cover of A^C which becomes finer and finer as one approaches A.

Sketch of Proof of 3.3   First, extend f to a Lipschitz function on all of $\mathbf{R}^m$ (see [Federer, 1, 2.10.43]). Second, by Rademacher's Theorem 3.2, f is differentiable almost everywhere. Third, by Lusin's theorem [Federer, 2.3.5], there is a closed subset E of A such that Df is continuous on E and $\mathscr{L}^m(A - E) < \varepsilon$. Fourth, for any $a \in E$, $\delta > 0$, define

$$\eta_\delta(a) = \sup_{\substack{0 < |x-a| < \delta \\ x \in E}} \frac{|f(x) - f(a) - Df(a)(x - a)|}{|x - a|}.$$

Since as $\delta \to 0$, $\eta_\delta \to 0$ pointwise, then by Egoroff's theorem [Federer, 2.3.7] there is a closed subset F of E such that $\mathscr{L}^m(A - F) < \varepsilon$ and $\eta_\delta \to 0$ uniformly on compact subsets of F. This condition implies the hypotheses of Whitney's extension theorem (3.4), with $P_a(x) = f(a) + Df(a)(x - a)$. Consequently, there is a C^1 function $g : \mathbf{R}^m \to \mathbf{R}^n$ which coincides with f on F.   ∎

The following theorem implies, for example, that Lipschitz images of sets of Hausdorff measure 0 have measure 0.

3.5 Proposition [Federer, 2.10.11] *Suppose $f : \mathbf{R}^l \to \mathbf{R}^n$ is Lipschitz and A is a Borel subset of $\mathbf{R}^l$. Then*

$$\int_{R^n} N(f \,|\, A, \, y)\, d\,\mathscr{H}^m y \leq (\operatorname{Lip} f)^m \,\mathscr{H}^m (A).$$

Here, $N(f \,|\, A, \, y) \equiv \operatorname{card}\{x \in A : f(x) = y\}$.

Proof Any covering of A by sets S_i of diameter d_i yields a covering of $f(A)$ by the sets $f(S_i)$, of diameter at most $(\operatorname{Lip} f)d_i$. Since the approximating sum $\sum \alpha_m (\operatorname{diam}/2)^m$ for the Hausdorff measure contains $(\operatorname{diam})^m$,

$$\mathscr{H}^m (f(A)) \leq (\operatorname{Lip} f)^m \mathscr{H}^m (A).$$

Notice that this formula gives the proposition in the case that f is injective. In the general case, chop A up into little pieces A_i and add up the formulas for each piece to obtain

$$\int_{f(A)} (\text{the number of } A_i \text{ intersecting } f^{-1}\{y\})\, d\,\mathscr{H}^m y \leq (\operatorname{Lip} f)^m \,\mathscr{H}^m (A).$$

As the pieces subdivide, the integrand increases monotonically to the multiplicity function $N(f \,|\, A, y)$, and the proposition is proved. ∎

The beginning of this proof illustrates the virtue of allowing coverings by arbitrary sets rather than just balls in the definition of Hausdorff measure. If $\{S_i\}$ covers A, then $\{f(S_i)\}$ is an admissible covering of $f(A)$.

3.6 Jacobians Jacobians are the corrective factors relating the elements of areas of the domains and images of functions. If $f : \mathbf{R}^m \to \mathbf{R}^n$ is differentiable at a, we define the *k-dimensional Jacobian of f at a*, $J_k f(a)$, as the maximum k-dimensional volume of the image under $Df(a)$ of a unit k-dimensional cube. If rank $Df(a) < k$, $J_k f(a) = 0$. If rank $Df(a) \leq k$, as holds in most applications, then $J_k f(a)^2$ equals the sum of the squares of the determinants of the $k \times k$ submatrices of $Df(a)$. If $k = m$ or n, then $J_k f(a)^2$ equals the determinant of the $k \times k$ product of $Df(a)$ with its transpose. If $k = m = n$, then $J_k f(a)$ is just the absolute value of the determinant of $Df(a)$. In general, computations are sometimes simplified by viewing $Df(a)$ as a map from the orthogonal complement

of its kernel onto its image. If $L : \mathbf{R}^m \to \mathbf{R}^m$ is linear, then $\mathscr{L}^m(L(A)) = J_m L \cdot \mathscr{L}^m(A)$.

3.7   The Area Formula [Federer, 3.2.3]   *Consider a Lipschitz function* $f : \mathbf{R}^m \to \mathbf{R}^n$ *for* $m \leq n$.

(1) *If A is an* $\mathscr{L}^m$ *measurable set, then*

$$\int_A J_m f(x)\, d\mathscr{L}^m x = \int_{\mathbf{R}^n} N(f|A, y)\, d\mathscr{H}^m y.$$

(2) *If u is an* $\mathscr{L}^m$ *integrable function, then*

$$\int_{\mathbf{R}^m} u(x) J_m f(x)\, dL^m x = \int_{\mathbf{R}^n} \sum_{x \in f^{-1}\{y\}} u(x)\, d\mathscr{H}^m y.$$

Remark   If f is a smooth embedding, then (1) equates the classical area of the parameterized surface $f(A)$ with the Hausdorff measure of $f(A)$. Therefore, for all smooth surfaces, the Hausdorff measure coincides with the classical area.

Sketch of the Proof of the Area Formula 3.7(1)   We will split up A into two cases, according to the rank of Df. In either case, by Rademacher's Theorem 3.2 and 3.5, we may assume that f is differentiable.

CASE 1 Df has rank m. Let $\{s_i\}$ be a countable dense set of affine maps of $\mathbf{R}^m$ onto m-dimensional planes in $\mathbf{R}^n$. Let E_i be a piece of A such that for each $a \in E_i$ the affine functions $f(a) + Df(a)(x - a)$ and $s_i(x)$ are approximately equal. It follows that

(1) $\det s_i \approx J_m f$ on E_i,
(2) f is injective on E_i, and
(3) the associated map from $s_i(E_i)$ to $f(E_i)$ and its inverse both have Lipschitz constant ≈ 1.

Because f is differentiable, the E_i cover A. Refine $\{E_i\}$ into a countable disjoint covering of A by tiny pieces. On each piece E, by (3) and 3.5,

$$\mathscr{H}^m(f(E)) \approx \mathscr{H}^m(s_i(E)) = \mathscr{L}^m(s_i(E))$$

$$= \int_E \det s_i\, d\mathscr{L}^m \approx \int_E J_m f\, d\mathscr{L}^m.$$

Summing over all the sets E yields

$$\int (\text{number of sets } E \text{ intersecting } f^{-1}\{y\}) d\,\mathcal{H}^m y \approx \int_A J_m f \, d\mathcal{L}^m.$$

Taking a limit yields

$$\int N(f|A, \, y) d\,\mathcal{H}^m y = \int_A J_m f \, d\mathcal{L}^m$$

and completes the proof of Case 1.

We remark that it does not suffice in the proof just to cut A up into tiny pieces without using the s_i. Without the requirement that for $a, b \in E$, $Df(a) \approx Df(b)$, f need not even be injective on E, no matter how small E is.

CASE 2 *DF has rank $< m$.* In this case, the left-hand side $\int_A J_m f$ is zero. Define a function

$$g : \mathbf{R}^m \to \mathbf{R}^{n+m}$$

$$x \to (f(x), \varepsilon x).$$

Then $J_m(g) \leq \varepsilon(\text{Lip } f + \varepsilon)^{m-1}$. Now by Case 1,

$$\mathcal{H}^m(f(A)) \leq \mathcal{H}^m(g(A)) = \int_A J_m g$$

$$\leq \varepsilon(\text{Lip } f + \varepsilon)^{m-1}\mathcal{L}^m(A).$$

Therefore, the right-hand side also must vanish. Finally, we remark that 3.7(2) follows from 3.7(1) by approximating u by simple functions. ∎

The following useful formula relates integrals of a function f over a set A to the areas of the level sets $A \cap f^{-1}\{y\}$ of the function.

3.8 The Coarea Formula [Federer, 3.2.11] *Consider a Lipschitz function $f : \mathbf{R}^m \to \mathbf{R}^n$ with $m > n$. If A is an $\mathcal{L}^m$ measurable set, then*

$$\int_A J_n f(x) \, d\mathcal{L}^m x = \int_{\mathbf{R}^n} \mathcal{H}^{m-n}(A \cap f^{-1}\{y\}) \, d\mathcal{L}^n y.$$

Proof. **CASE 1** *f is orthogonal projection.* If f is orthogonal projection, then $J_n f = 1$, and the coarea formula is reduced to Fubini's theorem.

GENERAL CASE We treat just the main case $J_n f \neq 0$. By subdividing A as in the proof of the area formula, we may assume that f is linear. Then $f = L \circ P$, where P denotes projection onto the n-dimensional orthogonal complement V of the kernel of f and where L is a nonsingular linear map from V to $\mathbf{R}^n$. Now

$$\int_A J_n f \, d\mathcal{L}^m = |\det L| \mathcal{H}^m(A)$$

$$= |\det L| \int_{P(A)} \mathcal{H}^{m-n}(P^{-1}\{y\}) \, d\mathcal{L}^n y$$

$$= \int_{L \circ P(A)} \mathcal{H}^{m-n}((L \circ P)^{-1}\{y\}) \, d\mathcal{L}^n y$$

as desired.

3.9 Tangent Cones

Suppose that $a \in \mathbf{R}^n$, $E \subset \mathbf{R}^n$, and φ is a measure on $\mathbf{R}^n$. Define a measure $\varphi \llcorner E$, "the restriction of φ to E," by

$$(\varphi \llcorner E)(A) = \varphi(E \cap A).$$

As in 2.5, define m-dimensional densities [Federer, 2.10.19]

$$\Theta^m(\varphi, a) = \lim_{r \to 0} \frac{\varphi(\mathbf{B}(a, r))}{\alpha_m r^m},$$

$$\Theta^m(E, a) = \Theta^m(\mathcal{H}^m \llcorner E, a)$$

$$= \lim_{r \to 0} \frac{\mathcal{H}^m(E \cap \mathbf{B}(a, r))}{\alpha_m r^m}.$$

Define the *tangent cone* of E at a consisting of the *tangent vectors* of E at a:

$$\mathrm{Tan}(E, a) = \{r \in \mathbf{R}: r \geq 0\} \left[\bigcap_{\varepsilon > 0} \mathrm{Clos} \left\{ \frac{x - a}{|x - a|} : x \in E, 0 < |x - a| < \varepsilon \right\} \right]$$

[Federer, 3.1.21].

Define the (smaller) cone of approximate tangent vectors of E at a:

$$\mathrm{Tan}^m(E, a) = \bigcap \{\mathrm{Tan}(S, a): \Theta^m(E - S, a) = 0\}$$

[Federer, 3.2.16]. See Figure 3.9.1.

3.10 Rectifiable Sets [Federer, 3.2.14]

A set $E \subset \mathbf{R}^n$ is called $(\mathcal{H}^m, m)$ *rectifiable* if $\mathcal{H}^m(E) < \infty$ and $\mathcal{H}^m$ almost all of E is contained in the union of the images of countably many Lipschitz functions from $\mathbf{R}^m$ to $\mathbf{R}^n$. These sets are the generalized surfaces of geometric measure theory. They include

Figure 3.9.1. A set, its tangent cone, and its approximate tangent cone at a. The approximate tangent cone ignores lower-dimensional pieces.

Figure 3.10.1. A two-dimensional rectifiable set in $\mathbf{R}^3$ consisting of the surfaces of countably many bicycles (but not the fractal closure).

countable unions of immersed manifolds (as long as the total area stays finite) and arbitrary subsets of $\mathbf{R}^m$.

Rectifiable sets can have countably many rectifiable pieces, perhaps connected by countably many tubes and handles and perhaps with all points in $\mathbf{R}^n$ as limit points (cf. Figure 3.10.1). Nevertheless, we will see that from the point of view of measure theory, rectifiable sets behave like C^1 submanifolds.

This book will call an $(\mathscr{H}^m, m)$-rectifiable and $\mathscr{H}^m$-measurable set an *m-dimensional rectifiable set.*

The following proposition shows that a measurable set E is rectifiable if and only if $\mathscr{H}^m(E) < \infty$ and $\mathscr{H}^m$ almost all of E is contained in a countable union of C^1, embedded manifolds.

3.11 Proposition [cf. Federer 3.2.18] *In the definition of a rectifiable set E, one can take the Lipschitz functions to be C^1 diffeomorphisms f_j*

on compact domains with disjoint images whose union coincides with E $\mathcal{H}^m$ almost everywhere. Moreover, the Lipschitz constants of f_j and f_j^{-1} can be taken near 1.

Proof   It suffices to obtain 1% of the set; the rest can be exhausted by repetition. By subdividing them we may assume that the domains have diameter at most 1. The first Lipschitz function f can be replaced by a C^1 approximation g by Theorem 3.3. By the area formula 3.7, we may assume Dg is nonsingular. By subdividing the domain, we may assume it is reasonably small. Take just a portion of the domain so that image $g \subset$ image E, Dg is approximately constant, and hence g is injective. Altering domain g by a linear transformation makes $Dg \approx$ identity, and Lip $g \approx$ Lip $g^{-1} \approx 1$. Finally, the domain may be replaced by a compact subset. Thus, 1% of image f is obtained. Similarly, replace the second Lipschitz function by a nice one with disjoint image. Continuing through all the original Lipschitz functions yields 1% of the set E. Countably many repetitions of the whole process prove the proposition.   ∎

The following proposition shows that in a certain sense a rectifiable set has a tangent plane at almost every point.

3.12   Proposition [Federer, 3.2.19]   *If W is an m-dimensional rectifiable subset of $\mathbf{R}^n$, then for almost all points a in W, the density $\Theta^m(W, a) = 1$ and $Tan^m(W, a)$ is an m-dimensional plane. If f is a Lipschitz map from W to $\mathbf{R}^v$, then f is approximately differentiable $\mathcal{H}^m$ almost everywhere.*

Remark   Preiss (see also De Lellis) proved a strong converse—that a Borel set of locally finite Hausdorff measure is rectifiable if the density exists and is positive at almost all points.

Example   This example gives a modest indication of how bad rectifiable sets can be and hence the strength of Proposition 3.12. Begin by constructing a Cantor-like set of positive measure as follows. Start with the unit interval. First, remove the middle open interval of length $1/4$. (See Figure 3.12.1.) Second, from the two remaining intervals, remove middle open intervals of total length $1/8$. At the nth step, from the 2^{n-1} remaining intervals, remove middle open intervals of total length $2^{-(n+1)}$. Let C be the intersection. Clearly, C contains no interval. However, since the total length removed was $\sum 2^{-(n+1)} = 1/2$, the length remaining $\mathcal{H}^1(C) = 1/2$.

Figure 3.12.1.   A Cantor-like set C with $\mathcal{H}^1(C) = 1/2$.

Figure 3.12.2. The image of g intersects $[0, 1]$ in the set C.

Now define $g : [0, 1] \to \mathbf{R}^2$ by

$$g(x) = (x, \text{dist}(x, C)).$$

See Figure 3.12.2.

Then image g and hence $E = [0, 1] \cup (\text{image } g)$ are rectifiable, even though E fails to be a submanifold at all points of C. Nevertheless, Proposition 3.12 states that $\Theta^1(E, x) = 1$ and $\text{Tan}^1(E, x)$ is a line at almost all points $x \in C$.

Remarks on Proof The proof that $\Theta^m(W, a) = 1$ almost everywhere uses a covering argument (see Corollary 2.9).

Proposition 3.11 implies that $\text{Tan}^m(W, a)$ contains an m-plane almost everywhere. Since $\Theta^m(W, a) = 1$, it can contain no more.

Similarly by Proposition 3.11, at almost every point, neglecting sets of density 0, W is parameterized by a nonsingular C^1 map $g : \mathbf{R}^m \to \mathbf{R}^n$. By Rademacher's theorem, 3.2, $f \circ g$ is differentiable almost everywhere, and hence f is approximately differentiable almost everywhere.

Here, we state a general theorem which subsumes both the area and the coarea formula.

3.13 General Area–Coarea Formula [Federer, 3.2.22] *Let W be an m-dimensional rectifiable subset of $\mathbf{R}^n$, Z a μ-dimensional rectifiable subset of $\mathbf{R}^v$, $m \geq \mu \geq 1$, and f a Lipschitz function from W to Z. Then*

$$\int_W a \, p \, J_\mu f \, d\mathcal{H}^m = \int_Z \mathcal{H}^{m-\mu}(f^{-1}\{z\}) \, d\,\mathcal{H}^\mu z.$$

More generally, for any $\mathcal{H}^m \llcorner W$ integrable function g on W,

$$\int_W g \cdot a \, p \, J_\mu f \, d\mathcal{H}^m = \int_Z \int_{f^{-1}\{z\}} g \, d\,\mathcal{H}^{m-\mu} \, d\,\mathcal{H}^\mu z.$$

Note: If f has an extension $\bar{f}$ to $\mathbf{R}^n$, $a \, p J_\mu f \leq J_\mu \bar{f}$ (where both are defined).

3.14 Product of Measures [Federer, 3.2.23] *Let W be an m-dimensional rectifiable Borel subset of $\mathbf{R}^n$ and let Z be a μ-dimensional rectifiable Borel*

subset of $\mathbf{R}^v$. *If W is contained in the image of a single Lipschitz function on a bounded subset of* $\mathbf{R}^m$, *then* $W \times Z$ *is rectifiable and*

$$\mathscr{H}^{m+\mu} \llcorner (W \times Z) = (\mathscr{H}^m \llcorner W) \times (\mathscr{H}^\mu \llcorner Z).$$

Remarks In general, the additional hypothesis on W is necessary. If $\mu = v$, it holds automatically. In particular, if Z is a μ-dimensional rectifiable Borel subset of $\mathbf{R}^v$, then $[0, 1]^m \times Z$ is an $(m + \mu)$-dimensional rectifiable subset of $\mathbf{R}^{m+v}$. If $m = n$ and $\mu = v$, this proposition is just Fubini's theorem.

The proof, as that of Fubini's theorem, shows that the collection of sets on which the proposition holds is a σ-algebra.

3.15 Orientation An *orientation* of an m-dimensional rectifiable subset W of $\mathbf{R}^n$ is a (measurable) choice of orientation for each $\mathrm{Tan}^m(W, a)$. At present, no further coherence is required, but we will see in Section 4.2 that a bad choice will make the boundary ∂W much worse. Every rectifiable set of positive measure has uncountably many different orientations (not just two).

3.16 Crofton's Formula [Federer, 3.2.26] *If W is an m-dimensional rectifiable set, then the integral-geometric measure of W equals its Hausdorff measure:*

$$\mathscr{I}^m(W) = \mathscr{H}^m(W).$$

Remarks Crofton's formula follows easily from the coarea formula. The proof, although stated for one-dimensional measure in $\mathbf{R}^2$, applies virtually unchanged to m-dimensional measure in $\mathbf{R}^n$.

Proof For a one-dimensional measure in $\mathbf{R}^2$,

$$\mathscr{H}^1(W) = \int_W (\text{length of unit tangent}) \, d\mathscr{H}^1$$

$$= \int_W \frac{1}{\beta(2, 1)} \int_{p \in \mathbf{O}^*(2, 1)} (\text{length of projection of unit tangent}) \, dp \, d\mathscr{H}^1$$

[because $\mathscr{I}^1$ (unit tangent) $= 1$]

$$= \frac{1}{\beta(2, 1)} \int_{p \in \mathbf{O}^*(2, 1)} \int_W (\text{length of projection of unit tangent}) \, d\mathscr{H}^1 \, dp$$

$$= \frac{1}{\beta(2, 1)} \int_{p \in \mathbf{O}^*(2, 1)} \int_W J_1 p \, d\mathscr{H}^1 \, dp$$

$$= \frac{1}{\beta(2,\,1)} \int_{p \in \mathbf{O}^*(2,\,1)} \int N(p|W,\,y)\, d\,\mathscr{H}^1\, y\, dp$$

(by the coarea formula, 3.13, because W rectifiable)

$$= \mathscr{I}^1(W).$$

The proof is virtually identical in general dimensions.   ■

3.17 Structure Theorem [Federer, 3.3.13] This striking theorem describes the structure of arbitrary subsets of $\mathbf{R}^n$. Proved for one-dimensional subsets of $\mathbf{R}^2$ by Besicovitch in 1939, it was generalized to general dimensions by Federer in 1947.

Let E be an arbitrary subset of $\mathbf{R}^n$ with $\mathscr{H}^m(E) < \infty$. Then E can be decomposed as the union of two disjoint sets $E = A \cup B$ with A $(\mathscr{H}^m, m)$ rectifiable and $\mathscr{I}^m(B) = 0$.

Remarks   That $\mathscr{I}^m = 0$ means that almost all of its projections onto m-planes have measure 0; we might say B is invisible from almost all directions. Such a set B is called *purely unrectifiable.*

The proof, a technical triumph, employs numerous ingenious coverings, notions of density, and amazing dichotomies. A nice presentation of Besicovitch's original proof of the structure theorem for one-dimensional subsets of the plane appears in [Falconer, Chapter 3].

Structure theory had been considered the most daunting component of the proof of the compactness theorem for integral currents, 5.5. In 1986, following Bruce Solomon, Brian White [3] found a direct argument that obviated the dependence on structure theory. In 1998, White [4] gave an easier proof by induction of the structure theorem.

If E is Borel, so are A and B.

Example   Purely unrectifiable sets result from Cantor-type constructions. For example, start with the unit square. Remove a central cross, leaving 4 squares, each 1/4 as long as the first. (See Figure 3.17.1.) Similarly, remove central crosses from each small square, leaving 16 smaller squares. Continue, and let the set E be the intersection.

The set E is purely unrectifiable. $\mathscr{H}^1(E) = \sqrt{2}$, but $\mathscr{I}^1(E) = 0$. Almost all projections onto lines have measure 0. For example, the projection onto the x-axis is itself a slim Cantor-like set of dimension 1/2. A diagonal line (with slope 1/2) gives an exceptional case: The projection is a solid interval. If A is any rectifiable set, then $\mathscr{H}^1(A \cap E) = 0$.

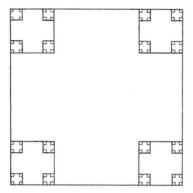

Figure 3.17.1. A purely unrectifiable one-dimensional set E. E is invisible from almost all directions.

Exercises

3.1 Give an example of a Lipschitz function $f : [0, 1] \to \mathbf{R}$ such that f is not differentiable at any rational point.

3.2 Use Theorem 3.3 to deduce that a Lipschitz function is approximately differentiable almost everywhere.

3.3 Give an example of a continuous function $f : \mathbf{R} \to \mathbf{R}$ such that

   a. given $\varepsilon > 0$ there is a C^1 function $g : \mathbf{R}^1 \to \mathbf{R}^1$ such that $\mathscr{L}^1\{x : f(x) \neq g(x)\} < \varepsilon$, but

   b. f is not Lipschitz.

3.4 Consider the map $f : \mathbf{R}^2 - \{0\} \to \mathbf{R}^2$ carrying Cartesian coordinates (x, y) to polar coordinates (r, θ). What is $J_1 f$?

3.5 Consider a differentiable map $f : \mathbf{R}^n \to \mathbf{R}$. Show that $J_1 f = |\nabla f|$.

3.6 Compute $\mathscr{H}^2$ of the unit two-sphere $\mathbf{S}^2(0, 1)$ by considering the map

$$f : \mathbf{R}^2 \to \mathbf{R}^3$$
$$f : (\varphi, \theta) \to (\sin\varphi \cos\theta, \ \sin\varphi \sin\theta, \ \cos\varphi).$$

3.7 Verify the coarea formula for $f : \mathbf{R}^3 \to \mathbf{R}$ given by $f(x, y, z) = x^2 + y^2 + z^2$, $A = \mathbf{B}^3(0, R)$.

3.8 Let E be an m-dimensional rectifiable Borel subset of the unit sphere in $\mathbf{R}^n$ with $\mathscr{H}^m(E) = a_0$. Let $C = \{tx : x \in E, 0 \le t \le 1\}$.

   a. Rigorously compute $\mathscr{H}^{m+1}(C)$.

   b. Compute $\Theta^{m+1}(C, 0)$.

   c. What is $\mathrm{Tan}^{m+1}(C, 0)$?

3.9 Give an example of an $(\mathscr{H}^2, 2)$-rectifiable subset E of $\mathbf{R}^3$ which is dense in $\mathbf{R}^3$. Can you also make $\{x \in \mathbf{R}^3 : \Theta^2(E, x) = 1\}$ dense in $\mathbf{R}^3$?

3.10 Let A be a compact subset of $\mathbf{R}^n$. Prove that the distance function $f(x) = \mathrm{dist}(x, A)$ is Lipschitz. Prove that f is differentiable at x outside A if and only if there is a unique closest point of A.

Normal and Rectifiable Currents

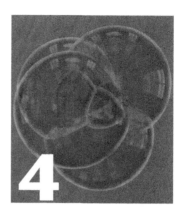

In order to define *boundary* and establish compactness properties, it will be useful to view our rectifiable sets as *currents*—that is, linear functionals on smooth differential forms (named by analogy with electrical currents with a kind of direction as well as magnitude at a point). The action of an oriented rectifiable set S on a differential form φ is given by integrating the form φ over the set:

$$S(\varphi) = \int_S \varphi \, d \, \mathcal{H}^m.$$

Currents thus associated with certain rectifiable sets, with integer multiplicities, will be called rectifiable currents. The larger class of normal currents will allow for real multiplicities and smoothing.

The concept of currents is a generalization, by de Rham [2], of distributions. Normal and rectifiable currents are due to Federer and Fleming. Important earlier and contemporaneous work includes the generalized surfaces of L. C. Young [1, 2], the frontiers of E. De Giorgi [1, 3, 4], and the surfaces of E. R. Reifenberg [1–3]. For hypersurfaces, rectifiable currents are just boundaries of the sets of finite perimeter of Caccioppoli and De Giorgi (see Giusti). The general reference for this chapter is [Federer, Chapter IV].

4.1 Vectors and Differential Forms [Federer, Chapter 1 and 4.1]

Consider $\mathbf{R}^n$ with basis $\mathbf{e}_1, \mathbf{e}_2, \ldots, \mathbf{e}_n$. There is a good way of multiplying m vectors in $\mathbf{R}^n$ to obtain a new object called an *m-vector* ξ:

$$\xi = v_1 \wedge \cdots \wedge v_m.$$

This wedge product is characterized by two properties. First, it is multilinear:

$$cv_1 \wedge v_2 = v_1 \wedge cv_2 = c(v_1 \wedge v_2),$$

$$(u_1 + v_1) \wedge (u_2 + v_2) = u_1 \wedge u_2 + u_1 \wedge v_2 + v_1 \wedge u_2 + v_1 \wedge v_2.$$

Geometric Measure Theory. http://dx.doi.org/10.1016/B978-0-12-804489-6.00004-6

Second, it is alternating:

$$u \wedge v = -v \wedge u \quad \text{or} \quad u \wedge u = 0.$$

For example,

$$(2\mathbf{e}_1 + 3\mathbf{e}_2 - 5\mathbf{e}_3) \wedge (7\mathbf{e}_1 - 11\mathbf{e}_3)$$
$$= 14\mathbf{e}_1 \wedge \mathbf{e}_1 - 22\mathbf{e}_1 \wedge \mathbf{e}_3 + 21\mathbf{e}_2 \wedge \mathbf{e}_1 - 33\mathbf{e}_2 \wedge \mathbf{e}_3$$
$$- 35\mathbf{e}_3 \wedge \mathbf{e}_1 + 55\mathbf{e}_3 \wedge \mathbf{e}_3$$
$$= 0 - 22\mathbf{e}_1 \wedge \mathbf{e}_3 - 21\mathbf{e}_1 \wedge \mathbf{e}_2 - 33\mathbf{e}_2 \wedge \mathbf{e}_3 + 35\mathbf{e}_1 \wedge \mathbf{e}_3 + 0$$
$$= -21\mathbf{e}_{12} + 13\mathbf{e}_{13} - 33\mathbf{e}_{23}.$$

We have abbreviated $\mathbf{e}_{12}$ for $\mathbf{e}_1 \wedge \mathbf{e}_2$.

In general, computation of $\xi = v_1 \wedge \cdots \wedge v_m$ yields an answer of the form

$$\xi = \sum_{i_1 < \cdots < i_m} a_{i_1 \cdots i_m} e_{i_1 \cdots i_m}.$$

The set of all linear combinations of $\{e_{i_1 \cdots i_m} : i_i < \cdots < i_m\}$ is the space $\Lambda_m \mathbf{R}^n$ of m-vectors, a vectorspace of dimension $\binom{n}{m}$. It has the inner product for which $\{e_{i_1 \cdots i_m}\}$ is an orthonormal basis.

The purpose of an m-vector $\xi = v_1 \wedge \cdots \wedge v_m$ is to represent the oriented m-plane P through $\mathbf{0}$ of which $v_l, \ldots, v_m$ give an oriented basis. Fortunately, the wedge product $\xi' = v'_1 \wedge \cdots \wedge v'_m$ of another oriented basis for P turns out to be a positive multiple of ξ. For example, replacing v_1 with $v'_1 = \sum c_i v_i$ yields

$$v'_1 \wedge v_2 \wedge \cdots \wedge v_m = c_1 v_1 \wedge v_2 \wedge \cdots \wedge v_m.$$

If $v_l, \ldots, v_m$ give an orthonormal basis, then $\xi = v_1 \wedge \cdots \wedge v_m$ has length 1. A product $v_1 \wedge \cdots \wedge v_m$ is 0 if and only if the vectors are linearly dependent. For the case $m = n$,

$$v_1 \wedge \cdots \wedge v_n = \det[v_1, \ldots, v_n] \cdot e_{1 \ldots n}.$$

An m-vector ξ is called *simple* or *decomposable* if it can be written as a single wedge product of vectors. For example, in $\Lambda_2 \mathbf{R}^4$, $\mathbf{e}_{12} + 2\mathbf{e}_{13} - \mathbf{e}_{23} = (\mathbf{e}_1 + \mathbf{e}_3) \wedge (\mathbf{e}_2 + 2\mathbf{e}_3)$ is simple, whereas $\mathbf{e}_{12} + \mathbf{e}_{34}$ is not (see Exercise 4.5). The oriented m-planes through the origin in $\mathbf{R}^n$ are in one-to-one correspondence with the unit, simple m-vectors in $\Lambda_m \mathbf{R}^n$.

Incidentally, the geometric relationship between two m-planes in $\mathbf{R}^n$ is given by m angles [Jordan], with beautiful later applications to the geometry of Grassmannians (see Wong) and to area minimization (see Morgan [1, §2.3]).

Let $\mathbf{R}^{n*}$ denote the space of *covectors* dual to $\mathbf{R}^n$, with dual orthonormal basis $\mathbf{e}_1^*, \ldots, \mathbf{e}_n^*$. We remark that dx_j is a common alternate notation for e_j^*. The dual space to $\Lambda_m \mathbf{R}^n$ is the space $\Lambda^m \mathbf{R}^n \equiv \Lambda_m(R^{n*})$ of linear combinations of wedge products of covectors, called m-covectors. The dual basis is $\{e^*_{i_1 \ldots i_m} : i_1 < \cdots < i_m\}$. A *differential m-form* φ on $\mathbf{R}^n$ is an m-covector field; that is, a map

$$\varphi : \mathbf{R}^n \to \Lambda^m \mathbf{R}^n.$$

For example, one 2-form on $\mathbf{R}^4$ is given by

$$\varphi = \cos x_1 e^*_{12} + \sin x_1 e^*_{34}$$
$$= \cos x_1 \, dx_1 \, dx_2 + \sin x_1 \, dx_3 \, dx_4.$$

The *support*, spt φ, of a differential form φ is defined as the closure of $\{x \in \mathbf{R}^n : \varphi(x) \neq 0\}$.

A differential m-form φ is a natural object to integrate over an oriented, m-dimensional rectifiable set S because it is sensitive to both the location $x \in S$ and the tangent plane to S at x. Let $\vec{S}(x)$ denote the unit m-vector associated with the oriented tangent plane to S at x. Then

$$\int_S \varphi \equiv \int_S \left\langle \vec{S}(x), \, \varphi(x) \right\rangle d\mathscr{H}^m x.$$

In a classical setting, with no Hausdorff measure available, the definition is more awkward. One uses local parameterizations and proves that the definition is independent of the choice of parameterization. Even the appropriateness of dealing with forms—functions on $\Lambda_m \mathbf{R}^n$—is obscured.

The *exterior derivative* $d\varphi$ of a differential m-form

$$\varphi = \sum f_{i_1 \ldots i_m} e^*_{i_1 \cdots i_m}$$

is the $(m + 1)$-form given by

$$d\varphi = \sum df_{i_1 \ldots i_m} \wedge e^*_{i_1 \cdots i_m},$$

where $df = (\partial f / \partial x_1) e^*_1 + \cdots + (\partial f / \partial x_n) e^*_n$. For example, if

$$\varphi = f \, dy \, dz + g \, dz \, dx + h \, dx \, dy,$$

then

$$d\varphi = \left(\frac{\partial f}{\partial x} + \frac{\partial g}{\partial y} + \frac{\partial h}{\partial z} \right) dx \, dy \, dz$$
$$= \operatorname{div}(f, g, h) \, dx \, dy \, dz.$$

If ϕ is a differential l-form and ω is a differential m-form, then

$$d(\phi \wedge \omega) = (d\phi) \wedge \omega + (-1)^l \phi \wedge d\omega.$$

In addition to the dual Euclidean norms $|\xi|$, $|\varphi|$ on $\Lambda_m \mathbf{R}^n$ and $\Lambda^m \mathbf{R}^n$, there are the mass norm $\|\xi\|$ and comass norm $\|\varphi\|^*$, also dual to each other, defined as follows:

$$\|\varphi\|^* = \sup\{|\langle \xi, \varphi \rangle| : \xi \text{ is a unit, simple } m\text{-vector}\};$$
$$\|\xi\| = \sup\{|\langle \xi, \varphi \rangle| : \|\varphi\|^* = 1\}.$$

It follows from convexity theory that

$$\|\xi\| = \inf\left\{\sum |\xi_i| : \xi = \sum \xi_i, \; \xi_i \text{ simple}\right\}.$$

Consequently, $\|\varphi\|^* = \sup\{|\langle \xi, \varphi \rangle| : \|\xi\| = 1\}$ so that the mass and comass norms are indeed dual to each other. Federer denotes both mass and comass norms by $\| \; \|$.

4.2 Currents [Federer, 4.1.1, 4.1.7]   The ambient space is $\mathbf{R}^n$. Let

$$\mathscr{D}^m = \{C^\infty \text{ differential } m\text{-forms with compact support}\}.$$

For example, in $\mathbf{R}^4$, a typical $\varphi \in \mathscr{D}^2$ takes the form

$$\begin{aligned}
\varphi &= f_1 \, dx_1 \, dx_2 + f_2 \, dx_1 \, dx_3 + f_3 \, dx_1 \, dx_4 + f_4 \, dx_2 \, dx_3 \\
&\quad + f_5 \, dx_2 \, dx_4 + f_6 \, dx_3 \, dx_4 \\
&= f_1 \, \mathbf{e}_{12}^* + f_2 \, \mathbf{e}_{13}^* + f_3 \, \mathbf{e}_{14}^* + f_4 \, \mathbf{e}_{23}^* + f_5 \, \mathbf{e}_{24}^* + f_6 \, \mathbf{e}_{34}^*,
\end{aligned}$$

where the f_j are C^∞ functions of compact support. The topology is generated by locally finite sets of conditions on the f_j and their derivatives of arbitrary order.

The dual space is denoted $\mathscr{D}_m$ and called the space of *m-dimensional currents*. This is a huge space. Under the *weak topology* (generally called the weak* topology) on $\mathscr{D}_m$, $T_j \to T$ if and only if $T_j(\varphi) \to T(\varphi)$ for all forms $\varphi \in \mathscr{D}^m$.

Any oriented m-dimensional rectifiable set may be viewed as a current as follows. Let $\vec{S}(x)$ denote the unit m-vector associated with the oriented tangent plane to S at x. Then for any differential m-form φ, define

$$S(\varphi) = \int_S \left\langle \vec{S}(x), \varphi \right\rangle d \, \mathscr{H}^m.$$

Furthermore, we will allow S to carry a positive integer multiplicity $\mu(x)$, with $\int_S \mu(x) \, d \, \mathscr{H}^m < \infty$, and define

$$S(\varphi) = \int_S \left\langle \vec{S}(x), \varphi \right\rangle \mu(x) \, d \, \mathscr{H}^m.$$

Finally, we will require that S have compact support. Such currents are called *rectifiable currents*.

Definitions for currents are by duality with forms. The boundary of an m-dimensional current $T \in \mathscr{D}_m$ is the $(m-1)$-dimensional current $\partial T \in \mathscr{D}_{m-1}$ defined by

$$\partial T(\varphi) = T(d\varphi).$$

By Stokes's theorem, this agrees with the usual definition of boundary if T is (integration over) a smooth oriented manifold with boundary. Notice that giving a piece of the manifold the opposite orientation would create additional boundary (as in Figure 4.3.6). A boundary has no boundary; that is, $\partial \circ \partial = 0$, as follows from the easy fact that $d \circ d = 0$.

The boundary of a rectifiable current S is generally not a rectifiable current. If it happens to be, then the original current S is called an *integral current*. The *support* of a current is the smallest closed set C such that

$$(\operatorname{spt}\varphi) \cap C = \varnothing \Rightarrow S(\varphi) = 0.$$

4.3 Important Spaces of Currents [Federer, 4.1.24, 4.1.22, 4.1.7, 4.1.5]

Figure 4.3.1 gives increasingly general spaces of currents that play an important role in geometric measure theory. Figure 4.3.2 shows some low-dimensional examples. The first tier has a polygonal curve, an integral current of finite length and finite boundary, and a rectifiable region of finite area but infinite boundary length, bounded by an integral flat chain of infinite length. The second tier allows real multiplicities and smoothing. The final rows illustrate more general currents without the same geometric significance.

Definitions   Let

$\mathscr{D}_m = \{m\text{-dimensional currents in }\mathbf{R}^n\}$,

$\mathscr{E}_m = \{T \in \mathscr{D}_m : \operatorname{spt} T \text{ is compact}\}$,

$\mathscr{R}_m = \{\text{rectifiable currents}\}$
$\quad = \{T \in \mathscr{E}_m \text{ associated with oriented rectifiable sets, with integer multiplicities, with finite total measure (counting multiplicities)}\}$,

$\mathscr{P}_m = \{\text{integral polyhedral chains}\}$
$\quad = \text{additive subgroup of } \mathscr{E}_m \text{ generated by classically oriented simplices,}$

$\mathbf{I}_m = \{\text{integral currents}\}$
$\quad = \{T \in \mathscr{R}_m : \partial T \in \mathscr{R}_{m-1}\}$,

$\mathscr{F}_m = \{\text{integral flat chains}\}$
$\quad = \{T + \partial S : T \in \mathscr{R}_m, \, S \in \mathscr{R}_{m+1}\}$.

The definitions of the second tier of spaces will appear in Section 4.5.

We also define two important seminorms (possibly infinite) on the space of currents $\mathscr{D}_m$: the mass $\mathbf{M}$ and the flat norm $\mathscr{F}$.

$\mathscr{I}_m$	$\subset$	$\mathbf{I}_m$	$\subset$	$\mathscr{R}_m$	$\subset$	$\mathscr{F}_m$
integral polyhedral chains		integral currents		rectifiable currents		integral flat chains
$\cap$		$\cap$		$\cap$		$\cap$
$\mathbf{P}_m$	$\subset$	$\mathbf{N}_m$	$\subset$	$\mathbf{R}_m$	$\subset$	$\mathbf{F}_m$
real polyhedral chains		normal currents				real flat chains
						$\cap$
						$\mathscr{E}_m \subset \mathscr{D}_m$

Figure 4.3.1. The increasingly general spaces of currents of geometric measure theory.

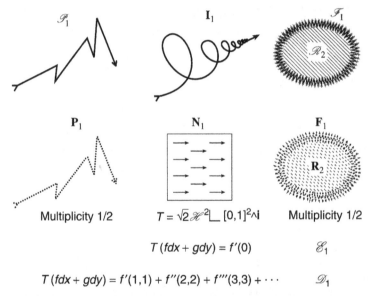

Multiplicity 1/2 $T = \sqrt{2}\,\mathscr{H}^2 \llcorner [0,1]^2 \wedge \mathbf{i}$ Multiplicity 1/2

$$T\,(f\,dx + g\,dy) = f'(0) \qquad \mathscr{E}_1$$

$$T\,(f\,dx + g\,dy) = f'(1,1) + f''(2,2) + f'''(3,3) + \cdots \qquad \mathscr{D}_1$$

Figure 4.3.2. Examples of increasingly general types of currents, with finite or infinite mass or boundary mass. The second tier admits fractional multiplicities and smoothing.

$$\mathbf{M}(T) = \sup\{T(\varphi) : \sup \|\varphi(x)\|^* \le 1\},$$

$$\mathscr{F}(T) = \inf\{\mathbf{M}(A) + \mathbf{M}(B) : T = A + \partial B,\ A \in \mathscr{R}_m,\ B \in \mathscr{R}_{m+1}\}.$$

The mass of a rectifiable current is just the Hausdorff measure of the associated rectifiable set (counting multiplicities) as explained in Section 4.5. Note that the norm $\sup_x \|\varphi(x)\|^*$ gives a weaker topology on $\mathscr{D}^m$ than the one to which currents

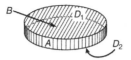

Figure 4.3.3. The unit discs D_1, D_2 are close together in the flat form $\mathscr{F}$ because their difference $T = D_2 - D_1$, together with a thin band A, is the boundary of a squat cylindrical region B of small volume. [Morgan 24, Figure 7].

are dual, so that a general current may well have infinite mass. Similarly, $\mathscr{F}(T) < \infty$ if and only if $T \in \mathscr{F}_m$.

The flat norm gives a good indication of when surfaces are geometrically close together. For example, the two (similarly oriented) unit discs D_1, D_2 of Figure 4.3.3 are close together under the flat norm $\mathscr{F}$ because their difference $T = D_2 - D_1$, together with a thin band A, is the boundary of a squat cylindrical region B of small volume. On the other hand, in the mass norm, $\mathbf{M}(D_2 - D_1) = 2\pi$.

The flat norm topology is clearly weaker than the mass norm topology but stronger than the weak topology. Actually, it turns out that for integral currents of bounded mass and boundary mass, the flat and weak topologies coincide as follows from the compactness theorem 5.5, or really just from Corollary 5.2 (compare Simon [3, 31.2]).

To obtain a rectifiable current which is not an integral current, choose the underlying rectifiable set E with infinite boundary. For example, let E be a connected open subset of the unit disc bounded by a curve of infinite length, as in Figure 4.3.4.

Alternatively, let E be a countable union of discs of radius $1/k$:

$$E = \bigcup_{k \in \mathbf{Z}^+} \{(x, y, z) : x^2 + y^2 \le k^{-2}, z = k^{-1}\}.$$

See Figure 4.3.5.

As a second alternative, decompose the unit disc into the infinitely many concentric annuli

$$A_n = \{1/(n+1) < r \le 1/n\}$$

of Figure 4.3.6 with alternating orientations.

In all three examples, the associated rectifiable current T is not an integral current, and ∂T is an integral flat chain but not a rectifiable current.

Actually, only by having infinite boundary mass can a rectifiable current fail to be an integral current. The difficult closure theorem, 5.4, will show that

(1)
$$\mathbf{I}_m = \{T \in \mathscr{R}_m : \mathbf{M}(\partial T) < \infty\},$$
$$\mathscr{R}_m = \{T \in \mathscr{F}_m : \mathbf{M}(T) < \infty\}.$$

(The equivalence of these two equalities follows immediately from the definitions.)

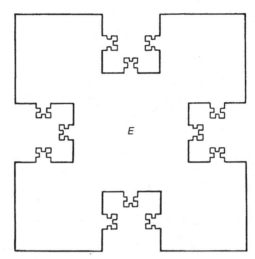

Figure 4.3.4. Although a rectifiable set *E* must have finite area, its boundary can wiggle enough to have infinite length. Thus, a rectifiable current need not be an integral current. Here, the width of each successive smaller square rapidly approaches one-third the length of the larger square.

Figure 4.3.5. This infinite collection of discs gives another example of a rectifiable current which is not an integral current. There is finite total area, but infinite total boundary length.

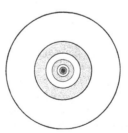

Figure 4.3.6. Giving alternating orientations to concentric annuli creates infinite boundary so that the disc is no longer an integral current. [Morgan 24, Figure 6].

Remarks on Supports and Notation   Let K be a compact C^1 submanifold of $\mathbf{R}^n$, with or without boundary (or more generally, a "compact Lipschitz neighborhood retract"). Federer uses the subscript K to denote support in K. For example,

$$\mathscr{R}_{m,K} = \{T \in \mathscr{R}_m : \mathrm{spt}\, T \subset K\}.$$

(For arbitrary compact K, $\mathscr{R}_{m,K}$ has a more technical meaning [Federer, 4.1.29].)

Similarly, a norm $\mathscr{F}_K$ is defined by

$$\mathscr{F}_K(T) \equiv \inf\{\mathbf{M}(A) + \mathbf{M}(B) : T = A + \partial B,\ A \in \mathscr{R}_{m,K},\ B \in \mathscr{R}_{m+1,K}\}.$$

If K is any large ball containing spt T, then $\mathscr{F}_K(T)$ equals what we have called $\mathscr{F}(T)$, as can be seen by projecting the A and B from the definition of F onto K. In the other main case of interest, when K is a compact C^1 submanifold of $\mathbf{R}^n$, $\mathscr{F}_K(T) \geq \mathscr{F}(T)$, with strict inequality sometimes. However, $\mathscr{F}_K$ and $\mathscr{F}$ yield the same topology on $\mathscr{F}_{m,K}$, the integral flat chains supported in K.

4.3A   Mapping Currents

Next we want to define the image of a compactly supported current under a proper C^∞ map $f : \mathbf{R}^n \to \mathbf{R}^v$. First, for any simple m-vector $\xi = v_1 \wedge \cdots \wedge v_m \in \wedge_m \mathbf{R}^n$, and point x in the domain of f, define the push-forward of ξ in $\wedge_m \mathbf{R}^v$:

$$[\wedge_m(Df(x))](\xi) = (Df(x))(v_1) \wedge \cdots \wedge (Df(x))(v_m).$$

The map $\wedge_m(Df(x))$ extends to a linear map on all m-vectors.

Second, for any differential m-form $\varphi \in \mathscr{D}^m(\mathbf{R}^v)$, define its pullback $f^\sharp \varphi$ on $\mathbf{R}^n$ by

$$\langle \xi,\ f^\sharp \varphi(x) \rangle = \langle [\wedge_m(Df(x))](\xi),\ \varphi(f(x)) \rangle.$$

Finally, for any compactly supported current $T \in \mathscr{D}_m(\mathbf{R}^n)$, define its push-forward $f_\sharp T \in \mathscr{D}_m(\mathbf{R}^v)$ by

$$(f_\sharp T)(\varphi) = T(f^\sharp \varphi).$$

If T is the rectifiable current associated with some oriented rectifiable set E, then $f_\sharp T$ is the rectifiable current associated with the oriented rectifiable set $f(E)$, with the appropriate multiplicities (see Exercise 4.23). The boundary $\partial(f_\sharp T) = f_\sharp \partial T$. In many cases, the smoothness hypothesis on f may be relaxed.

4.3B   Currents Representable by Integration

A current $T \in \mathscr{D}_m$ is called *representable by integration* if there are a Borel regular measure $\|T\|$ on

$\mathbf{R}^n$, finite on compact sets, and a function $\vec{T} : \mathbf{R}^n \to \Lambda_m \mathbf{R}^n$ with $\|\vec{T}(x)\| = 1$ for $\|T\|$ almost all x such that

$$T(\varphi) = \int \langle \vec{T}(x), \varphi(x) \rangle \, d \|T\| x.$$

The mass $\mathbf{M}(T)$ is just the total measure $\|T\| (\mathbf{R}^n)$. We write $T = \|T\| \wedge \vec{T}$.

A current $T \in \mathscr{D}_m$ of finite mass is automatically representable by integration, as follows from the representation theory of general measure theory. On the other hand, the current $T \in \mathscr{D}_m(\mathbf{R}^n)$ defined by

$$T(a_1 \, dx_1 \wedge \cdots \wedge dx_m + \cdots) = \frac{\partial a_1}{\partial x_1}(p),$$

where p is a fixed point in $\mathbf{R}^n$, has infinite mass and is not representable by integration.

Every rectifiable current S is representable by integration. Indeed, if E is the associated set with multiplicity function ℓ, then $\|S\|$ is the measure $\ell(\mathscr{H}^m \llcorner E)$ and $\vec{S}$ is the unit m-vector field orienting E. $S = \ell(\mathscr{H}^m \llcorner E) \wedge \vec{S} = (\mathscr{H}^m \llcorner E) \wedge \eta$, where $\eta = \ell \vec{S}$. The mass is

$$\mathbf{M}(S) = \|S\| (\mathbf{R}^n) = \int_S \ell \, d\mathscr{H}^m.$$

For example, the rectifiable current associated with a unit disc D in $\mathbf{R}^2 \subset \mathbf{R}^n$ is $\mathscr{H}^2 \llcorner D \wedge e_{12}$ and its mass is π.

4.4 Theorem [Federer, 4.1.28] *The following are equivalent definitions for $T \in \mathscr{E}_m$ to be a rectifiable current.*

(1) *Given $\varepsilon > 0$, there are an integral polyhedral chain $P \in \mathscr{P}_m(\mathbf{R}^v)$ and a Lipschitz function $f : \mathbf{R}^v \to \mathbf{R}^n$ such that*

$$\mathbf{M}(T - f_\sharp P) < \varepsilon.$$

(2) *There are a rectifiable set B and an $\mathscr{H}^m \llcorner B$ summable m-vector field η such that η is simple, $|\eta(x)|$ is an integer ("the multiplicity"), $\mathrm{Tan}^m(B, x)$ is associated with $\eta(x)$, and $T(\varphi) = \int_B \langle \eta(x), \varphi(x) \rangle \, d\mathscr{H}^m$.*

Remarks In (1), if T is supported in a closed ball K, one may assume spt $f_\sharp P \subset K$, by replacing $f_\sharp P$ by its projection onto K. Actually, Federer takes (1) as the definition of $\mathscr{R}_m$, whereas we have used (2).

A current $(\mathscr{H}^m \llcorner B) \wedge \eta$ can fail to be rectifiable in several ways: the set B could fail to be rectifiable or to have compact closure, the total mass $\int_B \|\eta(x)\| \, d\mathscr{H}^m$ could fail to be finite, the given m-vector $\eta(x)$ could fail to be tangent to B at x, or $|\eta(x)|$ could fail to be an integer.

Proof Sketch   First suppose (1) holds. Since each side of the polyhedral chain is a subset of some $\mathbf{R}^m$, its image under f is rectifiable and hence $f_\sharp P$ is a rectifiable current and satisfies (2). But now T, as a mass convergent sum of such, obtained by successive approximation, is a rectifiable current.

The opposite implication depends on the following lemma of measure theory.

Lemma 4.1   *Let A be a bounded ($\mathscr{L}^m$-measurable) subset of $\mathbf{R}^m$. Then given $\varepsilon > 0$, there is a finite disjoint set of m-simplices which coincide with A except for a set of measure less than ε.*

Proof of Lemma   We may assume that A is open, by replacing A by a slightly larger open set. Cover 1% of A by disjoint simplices (as in the proof of the Besicovitch covering theorem, 2.7). Repeat on what is left. After N repetitions, $1 - (.99)^N$ of A is covered by disjoint simplices, as desired.

Completion of Proof of Theorem   Suppose T satisfies (2). The rectifiable set B is the union of Lipschitz images of subsets of $\mathbf{R}^m$. Use the lemma to approximate B by images of polyhedra.   ∎

4.5   Normal Currents [Federer, 4.1.7, 4.1.12]   In preparation for the definitions of more general spaces of currents, we define a more general flat norm, $\mathbf{F}$. For any current $T \in \mathscr{D}_m$, define

$$\mathbf{F}(T) = \sup\{T(\varphi) : \varphi \in \mathscr{D}^m, \ \|\varphi(x)\|^* \leq 1 \text{ and } \|d\varphi(x)\|^* \leq 1 \text{ for all } x\}$$
$$= \min\{\mathbf{M}(A) + \mathbf{M}(B) : T = A + \partial B, \ A \in \mathscr{E}_m, \ B \in \mathscr{E}_{m+1}\}.$$

The second equality shows the similarity of the norm $\mathbf{F}$ and the previously defined norm $\mathscr{F}$. Inequality ($\leq$) is easy, since if $T = A + \partial B$ as in the minimum and φ is as in the supremum, then

$$T(\varphi) = (A + \partial B)(\varphi) = A(\varphi) + B(d\varphi) \leq \mathbf{M}(A) + \mathbf{M}(B).$$

Equality is proved using the Hahn–Banach theorem [Federer, p. 367].

Now continuing the definitions of the spaces of currents in the diagram in the beginning of Section 4.3, let

$$\mathbf{N}_m = \{T \in \mathscr{E}_m : \mathbf{M}(T) + \mathbf{M}(\partial T) < \infty\}$$
$$= \{T \in \mathscr{E}_m : T \text{ and } \partial T \text{ are representable by integration}\},$$
$$\mathbf{F}_m = \mathbf{F}\text{-closure of } \mathbf{N}_m \text{ in } \mathscr{E}_m,$$
$$\mathbf{R}_m = \{T \in \mathbf{F}_m : \mathbf{M}(T) < \infty\},$$
$$\mathbf{P}_m = \{\text{real linear combinations of elements of } \mathscr{P}_m\}.$$

Figure 4.5.1. Currents with noninteger densities and one-dimensional currents spread over two-dimensional sets give examples of normal currents which are not integral currents.

The important space $\mathbf{N}_m$ of normal currents allows real densities and smoothing. For example, if A is the unit square region

$$\{(x, y) : 0 \le x \le 1, 0 \le y \le 1\}$$

in the plane, then $S_1 = \sqrt{2}(\mathcal{H}^2 \llcorner A) \wedge \mathbf{e}_{12}$ is a two-dimensional normal current which is not an integral current. (See Figure 4.5.1.) S_1 is $\sqrt{2}$ times the integral current $(\mathcal{H}^2 \llcorner A) \wedge \mathbf{e}_{12}$. $S_2 = (\mathcal{H}^2 \llcorner A) \wedge \mathbf{e}_1$ is a one-dimensional normal current which is not an integral current. To check that S_2 is indeed a normal current, compute ∂S_2 from the definition:

$$\partial S_2(f(x, y)) = S_2(d f) = S_2 \left(\frac{\partial f}{\partial x} \mathbf{e}_1^* + \frac{\partial f}{\partial y} \mathbf{e}_2^* \right)$$

$$= \int_A \left\langle \mathbf{e}_1, \frac{\partial f}{\partial x} \mathbf{e}_1^* + \frac{\partial f}{\partial y} \mathbf{e}_2^* \right\rangle d\mathcal{H}^2$$

$$= \int_A \frac{\partial f}{\partial x} dx\, dy$$

$$= \int_0^1 [f(1, y) - f(0, y)]\, dy$$

$$= \int_0^1 f(1, y)\, dy - \int_0^1 f(0, y)\, dy.$$

Therefore,

$$\partial S_2 = \mathcal{H}^1 \llcorner \{(1, y) : 0 \le y \le 1\} - \mathcal{H}^1 \llcorner \{(0, y) : 0 \le y \le 1\},$$

and $\mathbf{M}(\partial S_2) = 2 < \infty$. If $B = \{(x, 0) : 0 \le x \le 1\}$, $T = \mathcal{H}^1 \llcorner B \wedge \mathbf{e}_1$, and $\tau_{(x, y)}$ denotes translation by (x, y), then

$$S_2 = \int_0^1 \tau_{(0, y)\sharp} T\, dy.$$

Thus, S_2 is an integral of integral currents.

More generally, if T is any m-dimensional integral current in $\mathbf{R}^n$ and f is a function of compact support with $\int |f| d\mathscr{L}^n < \infty$, then the weighted smoothing of T

$$S = \int_{x \in \mathbf{R}^n} f(x) \cdot \tau_{x\sharp} T \, d\mathscr{L}^n x$$

is a normal current. Of course,

$$\partial S = \int_{x \in \mathbf{R}^n} f(x) \cdot \tau_{x\sharp} \partial T \, d\mathscr{L}^n x.$$

Whether every normal current can be written as an integral of integral currents without cancellation has been a subject of research. A counterexample was provided by M. Zworski. (Incidentally, Zworski's theorem 2 needs an additional hypothesis; for example, a mass decomposition of $\partial \mathbf{N}$, or merely that the $\mathbf{R}_\omega$ have finite boundary mass, as G. Alberti pointed out to me.)

4.6 Proposition [Federer, 4.1.17]   *The space* $\mathbf{R}_m$ *is the* M-*closure of* $\mathbf{N}_m$ *in* $\mathscr{E}_m$.

Proof   Clearly, $\mathbf{R}_m$ is M-closed in $\mathscr{E}_m$. Suppose $T \in \mathbf{R}_m$. Given $\varepsilon > 0$, choose $S \in \mathbf{N}_m$ such that $\mathbf{F}(T - S) < \varepsilon$. Hence, there are currents $A \in \mathscr{E}_m$ and $B \in \mathscr{E}_{m+1}$ such that $T - S = A + \partial B$ and $\mathbf{M}(A) + \mathbf{M}(B) < \varepsilon$. Since $\mathbf{M}(\partial B) = \mathbf{M}(T - S - A) < \infty$, $\partial B \in \mathbf{N}_m$. Therefore, $S + \partial B \in \mathbf{N}_m$, and $\mathbf{M}(T - (S + \partial B)) = \mathbf{M}(A) < \varepsilon$. Hence, T is the M-closure of $\mathbf{N}_m$, as desired.   ∎

We have seen examples of m-dimensional normal currents based on higher-dimensional sets. The following theorem shows that even real flat chains cannot be supported in lower-dimensional sets. The hypothesis that the integral-geometric measure $\mathscr{I}^m (\operatorname{spt} T) = 0$ holds if the Hausdorff measure $\mathscr{H}^m (\operatorname{spt} T) = 0$, as follows easily from the definition of $\mathscr{I}^m$ (2.4).

4.7 Theorem [Federer, 4.1.20]   *If* $T \in \mathbf{F}_m(\mathbf{R}^n)$ *and* $\mathscr{I}^m (spt \, T) = 0$, *then* $T = 0$.

Examples   The current $S \equiv \mathscr{H}^0 \llcorner \{(0, 0)\} \wedge \mathbf{e}_1 \in \mathscr{D}_1$ is not flat because $\mathscr{I}^1 (\operatorname{spt} S) = \mathscr{H}^1 \{(0, 0)\} = 0$. The current

$$T = \mathscr{H}^1 \llcorner \{(0, y) : 0 \leq y \leq 1\} \wedge \mathbf{e}_1$$

is not flat, because if it were, its projection on the x-axis, which is S, would be flat. (See Figure 4.7.1.) This example illustrates the principle that for a flat current, the prescribed vector field must lie down "flat" (see Federer [1, 4.1.15]).

Figure 4.7.1. The current T is not flat; its prescribed vector field is not tangent to the underlying set.

The suggestiveness of the term *flat* is a happy accident. H. Whitney, also a student of music, coined the term for the smaller of his flat and sharp norms, originally designated $\| \ \|_\flat$ and $\| \ \|_\sharp$.

Outline of Proof

I. *Smoothing.* A smooth normal current in $\mathbf{R}^n$ is one of the form $\mathscr{L}^n \wedge \xi$, with ξ a smooth m-vector field of compact support. Any normal current T can be approximated in the flat norm by a smooth normal current $T_\varepsilon = \mathscr{L}^n \wedge \xi$ as follows. Let f be a smooth approximation to the delta function at 0, and put $T_\varepsilon = \int_{x \in \mathbf{R}^n} f(x) \cdot \tau_{x\sharp} T \, d\mathscr{L}^n x$.

II. *If $T \in \mathbf{F}_n(\mathbf{R}^n)$, then T is of the form $\mathscr{L}^n \wedge \xi$ for some vector field ξ* [Federer, 4.1.18]. Notice the assumption of codimension 0, where the norms $\mathbf{F}$ and $\mathbf{M}$ coincide. Therefore, T can be $\mathbf{M}$-approximated by a normal current and hence by smoothing by $\mathscr{L}^n \wedge \xi_1$, with ξ_1 a smooth n-vector field, $\mathbf{M}(T - \mathscr{L}^n \wedge \xi_1) < 2^{-1}$, and hence

$$\mathbf{M}(\mathscr{L}^n \wedge \xi_1) = \int |\xi_1| d\mathscr{L}^n < \mathbf{M}(T) + 2^{-1}.$$

Likewise, $T - \mathscr{L}^n \wedge \xi_1$ can be $\mathbf{M}$-approximated by $\mathscr{L}^n \wedge \xi_2$, with $\mathbf{M}(T - \mathscr{L}^n \wedge \xi_1 - \mathscr{L}^n \wedge \xi_2) < 2^{-2}$, and hence

$$\mathbf{M}(\mathscr{L}^n \wedge \xi_2) = \int |\xi_2| d\mathscr{L}^n < 2^{-1} + 2^{-2}.$$

Likewise, $T - \mathscr{L}^n \wedge \xi_1 - \mathscr{L}^n \wedge \xi_2$ can be M-approximated by $\mathscr{L}^n \wedge \xi_3$, with $\mathbf{M}(T - \mathscr{L}^n \wedge \xi_1 - \mathscr{L}^n \wedge \xi_2 - \mathscr{L}^n \wedge \xi_3) < 2^{-3}$, and hence $\mathbf{M}(\mathscr{L}^n \wedge \xi_3) < 2^{-2} + 2^{-3}$. Continue. Since $\int \sum_{j=1}^{\infty} |\xi_j| < \mathbf{M}(T) + 2^{-1} + 2^{-1} + 2^{-2} + 2^{-2} + \cdots = \mathbf{M}(T) + 2 < \infty$, $\sum \xi_j$ converges in L^1. Let $\xi = \sum \xi_j$. Then $T = \mathscr{L}^n \wedge \xi$ as desired.

III. *Completion of Proof.* For the case $m = n$, the theorem follows immediately from part II. Let $m < n$. Since $\mathscr{I}^m(\text{spt } T) = 0$, we may assume spt T projects to sets of measure 0 in the m-dimensional coordinate axis planes. For notational convenience we take $m = 1$ so that $T \in \mathbf{F}_1(\mathbf{R}^n)$. We consider the action of T on an arbitrary smooth 1-form

$$\varphi = f_1 \mathbf{e}_1^* + f_2 \mathbf{e}_2^* + \cdots + f_n \mathbf{e}_n^*.$$

Since $T(\varphi) = \sum T(f_j \mathbf{e}_j^*)$, it suffices to show that $T(f_j \mathbf{e}_j^*) = 0$. Let p_j denote projection onto the jth coordinate axis, and let $T \llcorner f$ denote the current defined by $(T \llcorner f)(\varphi) = T(f\varphi)$ (see 4.11). Then

$$T(f_j \mathbf{e}_j^*) = (T \llcorner f_j)(\mathbf{e}_j^*) = (T \llcorner f_j)(p_j^\sharp \mathbf{e}_j^*) = (p_{j\sharp}(T \llcorner f_j))(\mathbf{e}_j^*).$$

Since $p_{j\sharp}(T \llcorner f_j) \in \mathbf{F}_m(\mathbf{R}^m)$ is of the form $\mathscr{I}^m \wedge \xi$ by part II, and its support has measure 0, it must be 0. Therefore, $T(f_j \mathbf{e}_j^*) = 0$, as desired. ∎

4.8 Theorem [Federer, 4.1.23] *Given a real flat chain $T \in \mathbf{F}_m$ and $\varepsilon > 0$, there is a real polyhedral approximation $P \in \mathbf{P}_m$ satisfying $\mathbf{F}(T - P) \leq \varepsilon$ and $\mathbf{M}(P) \leq \mathbf{M}(T) + \varepsilon$.*

Proof Since the space $\mathbf{F}_m$ is defined as the F-closure of $\mathbf{N}_m$, and if $\mathbf{M}(T) < \infty$, T lies in the M-closure of $\mathbf{N}_m$ (Proposition 4.6), we may assume $T \in \mathbf{N}_m$. By smoothing (cf. proof of Theorem 4.7, part I), we may assume T is of the form $T = \mathscr{L}^n \wedge \xi(x)$, where $\xi(x)$ is a smooth m-vector field of compact support with $\int |\xi(x)| d\mathscr{L}^n < \infty$. By approximating ξ by step functions, we may assume T is of the form $T = \mathscr{L}^n \llcorner A \wedge \eta$, for some bounded set A and m-vector η. We may assume $\eta = \mathbf{e}_{1\cdots m}$ and A is the unit cube $\{0 \leq x_i \leq 1\} \subset \mathbf{R}^n$. Now we can approximate $T = \mathscr{L}^n \llcorner A \wedge \eta$ by layers. Take a large integer, M, let

$$B = \{x \in \mathbf{R}^m : 0 \leq x_i \leq 1\} \times \left\{\frac{1}{M}, \frac{2}{M}, \ldots, 1\right\}^{n-m} \subset \mathbf{R}^n,$$

and let

$$P = M^{-(n-m)}(\mathscr{H}^m \llcorner B) \wedge \mathbf{e}_{1\cdots m}.$$

Then $\mathbf{M}(P) = \mathbf{M}(T)$ and for M large, $\mathbf{F}(T - P) < \varepsilon$. ∎

4.9 Constancy Theorem [Federer, 4.1.31] *Suppose B is an m-dimensional connected, C^1 submanifold with boundary of $\mathbf{R}^n$, classically oriented by ζ. If a real flat chain $T \in \mathbf{F}_m$ is supported in B and its boundary is supported in the boundary of B, then, for some real number r,*

$$T = r(\mathscr{H}^m \llcorner B) \wedge \zeta.$$

Of course, if T is an integral flat chain, then r is an integer.

Proof   We must show locally that $\partial T = 0$ means T is constant. We may assume locally that $B = \mathbf{R}^m \times \{0\} \subset \mathbf{R}^n$. Then T is of the form $\mathscr{L}^m \wedge \xi$ for some m-vector field $\xi = f \cdot \mathbf{e}_{1 \cdots m}$ (proof of 4.7, part II). For any smooth $(m-1)$-form

$$\varphi = g_1 \mathbf{e}^*_{2 \cdots m} - g_2 \mathbf{e}^*_{13 \cdots m} + \cdots g_m \mathbf{e}^*_{12 \cdots m-1}$$

of compact support,

$$0 = \partial T(\varphi) = T(d\varphi) = \int \langle \xi, d\varphi \rangle \, d\mathscr{L}^m$$

$$= \int f \left(\frac{\partial g_1}{\partial x_1} + \frac{\partial g_2}{\partial x_2} + \cdots + \frac{\partial g_m}{\partial x_m} \right) d\mathscr{L}^m$$

$$= \int f \operatorname{div} g \, d\mathscr{L}^m.$$

It follows that f is constant, as desired. (If f is smooth, integration by parts yields that

$$0 = -\int \left(\frac{\partial f}{\partial x_1} g_1 + \cdots + \frac{\partial f}{\partial x_m} g_m \right) d\mathscr{L}^m$$

for all g_j so that $\partial f / \partial x_i = 0$ and f is constant. For general $f \in L^1$, $\int f \operatorname{div} g = 0$ for all g means the weak derivative vanishes and f is constant.) ∎

4.10 Cartesian Products   Given $S \in \mathscr{D}_m(\mathbf{R}^n)$ and $T \in \mathscr{D}_\mu(\mathbf{R}^v)$, one can define their Cartesian product $S \times T \in \mathscr{D}_{m+\mu}(\mathbf{R}^{n+v})$. The details appear in Federer [1, 4.1.8, p. 360], but for now it is enough to know that it exists and has the expected properties.

4.11 Slicing [Federer, 4.2.1]   The coarea formula 3.13 relates the area of a rectifiable set W to areas of its slices. In this section, we define $(m-1)$-dimensional slices of m-dimensional normal currents by hyperplanes or by hypersurfaces $\{u(x) = r\}$. It will turn out that for almost all values of r, the slices

themselves are normal currents, and that the boundary of the slice is just the slice of the boundary. For rectifiable sets the two notions of slicing agree (4.13).

First, for any current $T \in \mathscr{D}_m$ and C^∞ differential k-form α, define a current $T \llcorner \alpha \in \mathscr{D}_{m-k}$ by

$$(T \llcorner \alpha)(\varphi) = T(\alpha \wedge \varphi).$$

In particular, if α is a function (0-form) f, then $(T \llcorner f)(\varphi) = T(f\varphi)$. The symbol $\llcorner$ for such "interior multiplication," sometimes called "elbow," points to the term of lower degree which gets pushed to the other side in the definition.

If T is representable by integration, $T = \|T\| \wedge \vec{T}$, then it suffices to assume that $\int |f| d\|T\| < \infty$. Indeed, then $T \llcorner f = f \|T\| \wedge \vec{T}$: one just multiplies the multiplicity by f. Of course, even if T is rectifiable, $T \llcorner f$ will not be, unless f is integer valued. For $A \subset \mathbf{R}^n$, define "T restricted to A," $T \llcorner A = T \llcorner \chi_A$, where χ_A is the characteristic function of A.

For a normal current $T \in \mathbf{N}_m \mathbf{R}^n$, a Lipschitz function $u : \mathbf{R}^n \to \mathbf{R}$, and a real number r, define the slice

(1) $\langle T, u, r+ \rangle \equiv (\partial T) \llcorner \{x : u(x) > r\} - \partial(T \llcorner \{x : u(x) > r\})$

$\qquad = \partial(T \llcorner \{x : u(x) \le r\}) - (\partial T) \llcorner \{x : u(x) \le r\}.$

(See Figure 4.11.1.) It follows that

(2) $\qquad\qquad \partial \langle T, u, r+ \rangle = - \langle \partial T, u, r+ \rangle.$

Proposition

(3) $\qquad \mathbf{M} \langle T, u, r+ \rangle \le (\operatorname{Lip} u) \varliminf_{h \to 0+} \|T\| \{r < u(x) < r + h\}/h.$

In particular, if $f(r) = \|T\| B(x, r)$, then for almost all r,

$$\mathbf{M} \langle T, u, r+ \rangle \le f'(r).$$

Proof If χ is the characteristic function of the set $\{x : u(x) > r\}$, then

$$\langle T, u, r+ \rangle = (\partial T) \llcorner \chi - \partial(T \llcorner \chi).$$

For small, positive h, approximate χ by a C^∞ function f satisfying

$$f(x) = \begin{cases} 0 & \text{if } u(x) \le r \\ 1 & \text{if } u(x) \ge r + h \end{cases}$$

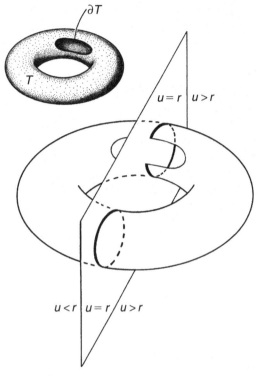

Figure 4.11.1. The slice of the torus T by the pictured plane consists of $1\frac{3}{4}$ circles.

and Lip $f \lesssim (\text{Lip}\, u)/h$. Then

$$\mathbf{M}\langle T,\, u,\, r+\rangle \approx \mathbf{M}((\partial T)\llcorner f - \partial(T\llcorner f))$$

$$= \mathbf{M}(T\llcorner d\, f)$$

$$\leq (\text{Lip}\, f)\|T\|\{x : r < u(x) < r + h\}$$

$$\lesssim (\text{Lip}\, u)\|T\|\{x : r < u(x) < r + h\}/h.$$

Consequently,

$$\mathbf{M}\langle T,\, u,\, r+\rangle \leq (\text{Lip}\, u) \varliminf_{h\to 0} \|T\|\{x : r < u(x) < r + h\}/h. \qquad \blacksquare$$

Proposition

(4) $$\int_a^b \mathbf{M}\langle T,\, u,\, r+\rangle\, d\mathscr{L}^1 r \leq (\text{Lip}\, u)\|T\|\{x : a < u(x) < b\}.$$

Proof Consider the function $f(r) = ||T||\{u(x) < r\}$. Since f is monotonically increasing, its derivative, $f'(r)$, exists for almost all r. Hence,

$$(\text{Lip}\,u)||T||\{a < u(x) < b\} = (\text{Lip}\,u)(f(b) - \lim_{x \to a+} f(x))$$

$$\geq (\text{Lip}\,u) \int_a^b f'(r)\,dr \geq \int_a^b \mathbf{M}\langle T, u, r+\rangle\,dr$$

by (3). ∎

Corollary

$$(5) \qquad \qquad \langle T, u, r+\rangle \in \mathbf{N}_{m-1}$$

for almost all r.

The corollary follows directly from (4) and (2) (see Exercise 4.20). Of course, it follows that if T is rectifiable, so are almost all slices, because the slices by definition belong to $\mathscr{F}$.

Proposition

$$(6) \qquad \int_a^b \mathbf{F}[T\llcorner\{u(x) \leq r\}]d\,\mathscr{L}^1 r \leq [b - a + \text{Lip}\,u]\mathbf{F}(T).$$

For a hint on the proof, see Exercise 4.21.

Remarks If T is an integral current, so is almost every slice, as will follow from the closure theorem, 5.4(2, 3), and 4.11(5) (or as is shown directly in Simon [3, §28]). Slicing can be generalized to a vector-valued function $u : \mathbf{R}^n \to \mathbf{R}^l$ [Federer, 4.3].

The following lemma considers slices of T by the function $u(x) = |x - a|$. If T has no boundary, then

$$\langle T, u, r+\rangle \equiv \partial(T\llcorner\{x : u(x) \leq r\}) = \partial(T\llcorner\mathbf{B}(a, r)).$$

The lemma says that if almost all such "slices by spheres" are rectifiable, then T is rectifiable.

4.12   Lemma [Federer, 4.2.15]   *If T is a normal current without boundary and if, for each $a \in \mathbf{R}^n$, $\partial(T\llcorner\mathbf{B}(a, r))$ is rectifiable for almost all $r \in \mathbf{R}$, then T is rectifiable.*

Remarks This lemma is crucial to proving the closure and compactness theorems of Chapter 5. The proof of this lemma in Federer uses structure theory and a covering argument. In 1986, following the work of Solomon, White [3] discovered a way to circumvent the structure theory at almost no cost. In 2000, Ambrosio and Kirchheim, in a more general setting, developed a rectifiability criterion involving only 0-dimensional slices.

4.13 Proposition [Federer, 4.3.8, 3.2.22] *Let W be an m-dimensional rectifiable set in R^n and u a Lipschitz function from W to R^μ. Then for almost all $z \in R^\mu$, $W \cap u^{-1}\{z\}$ is rectifiable, and the associated current is the slice $\langle T, u, r+ \rangle$ of the current T associated to W.*

Exercises

4.1 Compute $(\mathbf{e}_1 + 2\mathbf{e}_2 + 3\mathbf{e}_3) \wedge (\mathbf{e}_1 + 2\mathbf{e}_2 - 3\mathbf{e}_3) \wedge \mathbf{e}_4$.

4.2 Consider the 2-plane P in $\mathbf{R}^4$ given by

$$P = \{(x_1, x_2, x_3, x_4) : x_1 + x_2 + x_3 = x_3 + x_4 = 0\}.$$

Find a nonorthogonal basis u, v and an orthonormal basis w, z for P. Verify by direct computation that $u \wedge v$ is a multiple of $w \wedge z$ and that $|w \wedge z| = 1$.

4.3 Verify by direct computation that

$$(\mathbf{e}_1 + 2\mathbf{e}_2 + 3\mathbf{e}_3) \wedge (\mathbf{e}_1 - \mathbf{e}_3) \wedge (\mathbf{e}_2 + \mathbf{e}_3) = \begin{vmatrix} 1 & 1 & 0 \\ 2 & 0 & 1 \\ 3 & -1 & 1 \end{vmatrix} \mathbf{e}_{123}.$$

4.4 Prove that the 2-vector $\mathbf{e}_{12} + 2\mathbf{e}_{13} + 2\mathbf{e}_{23}$ is simple.

4.5 Prove that $\mathbf{e}_{12} + \mathbf{e}_{34}$ is not simple.

4.6 Find the integral of the differential form

$$\varphi = x_1(\sin x_1 x_2)\mathbf{e}_{12}^* + e^{x_1 + x_2 + x_3}\mathbf{e}_{13}^* + \mathbf{e}_{23}^*$$

over $\{(x_1, x_2, x_3) \in \mathbf{R}^3 : 0 \le x_1 \le 1, 0 \le x_2 \le 1, x_3 = 0\}$ (with the usual upward orientation).

4.7 Find the integral of the differential form

$$\varphi = 2\mathbf{e}_{12}^* + 3\mathbf{e}_{13}^* + 5\mathbf{e}_{23}^*$$

over $\{(x_1, x_2, x_3) \in \mathbf{R}^3 : x_1 + x_2 + x_3 = 0 \text{ and } x_1^2 + x_2^2 + x_3^2 \le 1\}$.

4.8 Prove that the boundary operator ∂ maps $\mathbf{I}_m$ into $\mathbf{I}_{m-1}$ and $\mathscr{F}_m$ into $\mathscr{F}_{m-1}$. Also prove that spt $\partial T \subset$ spt T.

4.9 For the rectifiable currents $T \in \mathscr{R}_1(\mathbf{R}^2)$ and for C^∞ functions f, g, and h, compute formulas for $T(f\, dx + g\, dy)$ and $\partial T(h)$:

a. $T = \mathscr{H}^1 \llcorner \{(x, 0) : 0 \le x \le 1\} \wedge \mathbf{e}_1$.
b. $T = (\mathscr{H}^1 \llcorner \{(x, x) : 0 \le x \le 1\}) \wedge 3\sqrt{2}(\mathbf{e}_1 + \mathbf{e}_2)$.

4.10 Prove that $\mathbf{I}_m$ is $\mathbf{M}$ dense in $\mathscr{R}_m$ and $\mathscr{F}$ dense in $\mathscr{F}_m$.

4.11 Prove that $\{T \in \mathscr{R}_m : \operatorname{spt} T \subset \mathbf{B}(0, R)\}$ is $\mathbf{M}$ complete and that $\{T \in \mathscr{F}_m : \operatorname{spt} T \subset \mathbf{B}(0, R)\}$ is $\mathscr{F}$ complete.

4.12 Prove that ∂ carries $\mathbf{N}_m$ into $\mathbf{N}_{m-1}$ and $\mathbf{F}_m$ into $\mathbf{F}_{m-1}$.

4.13 Check this analogy to 4.3(1):

$$\mathbf{N}_m = \{T \in \mathbf{R}_m : \mathbf{M}(\partial T) < \infty\},$$

$$\mathbf{R}_m = \{T \in \mathbf{F}_m : \mathbf{M}(T) < \infty\}.$$

4.14 Prove that, in analogy with the definitions of $\mathbf{I}_m$ and $\mathscr{F}_m$,

$$\mathbf{N}_m = \{T \in \mathbf{R}_m : \partial T \in \mathbf{R}_{m-1}\},$$

$$\mathbf{F}_m = \{T + \partial S : T \in \mathbf{R}_m, \ S \in \mathbf{R}_{m+1}\}.$$

4.15 Prove that $\mathbf{I}_m \subset \mathbf{N}_m$, $\mathscr{R}_m \subset \mathbf{R}_m$, and $\mathscr{F}_m \subset \mathbf{F}_m$.

4.16 For the currents $T \in \mathscr{D}_1(\mathbf{R}^2)$ representable by integration and for C^∞ functions f, g, and h,

  i. write a formula for $T(f \, dx + g \, dy)$,

  ii. write a formula for $\partial T(h)$, and

  iii. give the smallest space of currents from the table at the beginning of Section 4.3 to which T belongs:

  a. $T = \sum_{k=1}^{\infty} \mathscr{H}^1 \llcorner \{(k^{-1}, \ y) : 0 \le y \le 2^{-k}\} \wedge \mathbf{j}$.

  b. $T = \mathscr{H}^2 \llcorner \{(x, \ y) : 0 \le x \le 1, \ 0 \le y \le 1\} \wedge \mathbf{i}$.

  c. $T = \mathscr{H}^1 \llcorner \{(x, \ 0) : 0 \le x \le 1\} \wedge \mathbf{j}$.

  d. $T = \mathscr{H}^0 \llcorner \{a\} \wedge \mathbf{i}$.

  e. $T = \mathscr{H}^2 \llcorner \{x, \ y\} : x^2 + y^2 \le 1\} \wedge \mathbf{i}$.

4.17 Let E be the modification of the Cantor set obtained by starting with the unit interval and removing 2^{n-1} middle intervals, each of length 4^{-n}, at the nth step ($n = 1, 2, 3, \ldots$).

  a. Show that $\mathscr{H}^1(E) = 1/2$.

  b. Show that $\mathscr{H}^1 \llcorner E \wedge \mathbf{i}$ is a rectifiable current but not an integral current.

4.18 Prove the second equality in 4.11(1).

4.19 Prove 4.11(2).

4.20 Deduce 4.11(5) from 4.11(4) and 4.11(2).

4.21 Prove 4.11(6).

  Hint: First show that, if $T = A + \partial B$ with T, A, and $B \in \mathbf{N}$, then

  $$T \llcorner \{u(x) \le r\} = A \llcorner \{u(x) \le r\} + \partial[B \llcorner \{u(x) \le r\}] - \langle B, u, r+ \rangle.$$

4.22 Prove that $\mathbf{M}$ is $\mathbf{F}$ lower semicontinuous on $\mathscr{D}_m$; that is, if $T_i, T \in \mathscr{D}_m$, and $T_i \xrightarrow{\ \mathbf{F}\ } T$, then $\mathbf{M}(T) \le \liminf \mathbf{M}(T_i)$.

  Hint: Work right from the definition of $\mathbf{M}$ in 4.3 and the first definition of $\mathbf{F}$ in 4.5.

4.23 Suppose f is a C^∞ map from $\mathbf{R}^n$ to $\mathbf{R}^v$ and $S = l(\mathcal{H}^m \llcorner E) \wedge \vec{S}$ is a rectifiable current represented by integration in terms of an underlying rectifiable set E and an integer-valued multiplicity function l.

  a. Assuming that f is injective, show that

$$f_\sharp S = l \circ f^{-1}(\mathcal{H}^m \llcorner f(E)) \wedge \frac{(\Lambda_m Df)(\vec{S})}{(\Lambda_m Df)(\vec{S})}.$$

  b. Without assuming that f is injective, show that

$$f_\sharp S = (\mathcal{H}^m \llcorner f(E)) \wedge \sum_{y=f(x)} l(x) \frac{(\Lambda_m Df(x))(\vec{S})}{|(\Lambda_m Df(x))(\vec{S})|}.$$

Hint: Use the definitions and the general coarea formula, 3.13.

The Compactness Theorem and the Existence of Area-Minimizing Surfaces

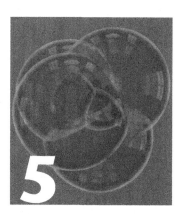

The compactness theorem, 5.5, deserves to be known as the fundamental theorem of geometric measure theory. It guarantees solutions to a wide class of variational problems in general dimensions. It states that a certain set $\mathscr{T}$ of surfaces is compact in a natural topology. The two main lemmas are the deformation theorem, 5.1, which will imply in 5.2 that $\mathscr{T}$ is totally bounded, and the closure theorem, 5.4, which will imply that $\mathscr{T}$ is complete.

5.1 The Deformation Theorem [Federer, 4.2.9]
The deformation theorem approximates an integral current (or similarly a normal current) T by deforming it onto a grid of mesh $2\varepsilon > 0$. (See Figure 5.1.1.) The resulting approximation, P, is automatically a polyhedral chain. The main error term is ∂S, where S is the surface through which T is deformed. There is a secondary error term, Q, due to moving ∂T into the skeleton of the grid.

Whenever $T \in \mathbf{I}_m \mathbf{R}^n$ *and* $\varepsilon > 0$, *there exist* $P \in \mathscr{P}_m \mathbf{R}^n$, $Q \in \mathbf{I}_m \mathbf{R}^n$, *and* $S \in \mathbf{I}_{m+1} \mathbf{R}^n$ *such that the following conditions hold with* $\gamma = 2n^{2m+2}$:

(1)
$$T = P + Q + \partial S.$$

(2)
$$\mathbf{M}(P) \leq \gamma [\mathbf{M}(T) + \varepsilon \mathbf{M}(\partial T)],$$

$$\mathbf{M}(\partial P) \leq \gamma \mathbf{M}(\partial T),$$

$$\mathbf{M}(Q) \leq \varepsilon \gamma \mathbf{M}(\partial T),$$

$$\mathbf{M}(S) \leq \varepsilon \gamma \mathbf{M}(T).$$

Consequently, $\mathscr{F}(T - P) \leq \varepsilon \gamma (\mathbf{M}(T) + \mathbf{M}(\partial T))$.

Geometric Measure Theory. http://dx.doi.org/10.1016/B978-0-12-804489-6.00005-8

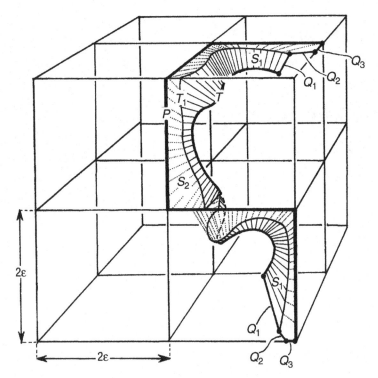

Figure 5.1.1. The deformation theorem describes a multistep process for deforming a given curve T onto a polygon P in the 2ε grid. During the process, surfaces S_1 and S_2 are swept out. The endpoints of T trace out curves Q_1, Q_2, and Q_3.

(3) spt P *is contained in the m-dimensional* 2ε *grid; that is, if* $x \in$ spt P, *then at least* $n - m$ *of its coordinates are even multiples of* ε. *Also,* spt ∂P *is contained in the* $(m - 1)$-*dimensional* 2ε *grid.*

(4)   spt P $\cup$ spt Q $\cup$ spt $S \subset \{x : \text{dist}(x, \text{ spt } T) \leq 2n\varepsilon\}$.

Proof Sketch, Case $m = 1, n = 3$. Let W_k denote the k-dimensional ε grid:

$$W_k = \{(x_1, x_2, x_3) \in \mathbf{R}^3 : \text{at least } 3 - k \text{ of the } x_j \text{ are even multiples of } \varepsilon\}.$$

Then W_2 consists of the boundaries of $2\varepsilon \times 2\varepsilon \times 2\varepsilon$ cubes.

First, project the curve T radially outward from the centers of the cubes onto W_2. (For now, suppose T stays away from the centers.) Let S_1 be the surface swept out by T during this projection, let Q_1 be the curve swept out by ∂T, and let T_1 be the image of T in W_2. Then, with suitable orientations,

$$T = T_1 + Q_1 + \partial S_1.$$

The mass of T_1 is of the same order as the mass of T, $\mathbf{M}(T_1) \sim \mathbf{M}(T)$. Likewise $\mathbf{M}(\partial T_1) \sim \mathbf{M}(\partial T)$, $\mathbf{M}(Q_1) \sim \varepsilon\mathbf{M}(\partial T)$, and $\mathbf{M}(S_1) \sim \varepsilon\mathbf{M}(T)$.

Second, W_1, the 1-skeleton of W_2, consists of the boundaries of $2\varepsilon \times 2\varepsilon$ square regions. Project the curve T_1 radially outward from the centers of the squares onto W_1. Let S_2 be the surface swept out by T_1, let Q_2 be the curve swept out by ∂T_1, and let T_2 be the image of T_1 in W_1. Then, $T_1 = T_2 + Q_2 + \partial S_2$, and the masses are of order $\mathbf{M}(T_2) \sim \mathbf{M}(T_1) \sim \mathbf{M}(T)$, $\mathbf{M}(\partial T_2) \sim \mathbf{M}(\partial T)$, $\mathbf{M}(Q_2) \sim \varepsilon\mathbf{M}(\partial T)$, and $\mathbf{M}(S_2) \sim \varepsilon\mathbf{M}(T)$.

Third, let Q_3 consist of line segments from each point of ∂T_2 to the nearest point in the 0-skeleton W_0. Put $P = T_2 - Q_3$. Then not only does P lie in W_1 but also ∂P lies in W_0. In particular, P is an integral polyhedral chain. The masses satisfy

$$\mathbf{M}(Q_3) \sim \varepsilon\mathbf{M}(\partial T),$$

$$\mathbf{M}(P) = \mathbf{M}(T_2) + \mathbf{M}(Q_3) \sim \mathbf{M}(T) + \varepsilon\mathbf{M}(\partial T),$$

$$\mathbf{M}(\partial P) = \mathbf{M}(\partial T_2) \sim \mathbf{M}(\partial T).$$

Let $Q = Q_1 + Q_2 + Q_3$ and $S = S_1 + S_2$. Then $T = P + Q + \partial S$. The masses satisfy

$$\mathbf{M}(Q) = \mathbf{M}(Q_1) + \mathbf{M}(Q_2) + \mathbf{M}(Q_3) \sim \varepsilon\mathbf{M}(\partial T),$$

$$\mathbf{M}(S) \sim \varepsilon\mathbf{M}(T),$$

completing the proof.

There is one problem with the foregoing sketch. If the original curve winds tightly about (or, worse, passes through) one of the centers of radial projection, the mass of its projection could be an order larger than its own mass. In this case, one moves the curve a bit before starting the whole process. If the original curve winds throughout space, it may be impossible to move it away from the centers of projection, but the average distortion can still be controlled. ∎

5.2 Corollary   *The set*

$$\mathscr{T} = \{T \in \mathbf{I}_m : \operatorname{spt} T \subset \mathbf{B}^n(\mathbf{0}, c_1),\ \mathbf{M}(T) \leq c_2,\ and\ \mathbf{M}(\partial T) \leq c_3\}$$

is totally bounded under $\mathscr{F}$.

Proof   Each $T \in \mathscr{T}$ can be well approximated by a polyhedral chain P in the ε-grid with $\mathbf{M}(P) \leq \gamma[c_2 + \varepsilon c_3]$ and $\operatorname{spt} P \subset \mathbf{B}^n(\mathbf{0}, c_1 + 2n\varepsilon)$ by 5.1(4). Since there are only finitely many such P, $\mathscr{T}$ is totally bounded. ∎

5.3 The Isoperimetric Inequality [Federer, 4.2.10] *If $T \in \mathbf{I}_m \mathbf{R}^n$ with $\partial T = 0$, then there exists $S \in \mathbf{I}_{m+1} \mathbf{R}^n$ with $\partial S = T$ and*

$$\mathbf{M}(S)^{m/(m+1)} \leq \gamma \mathbf{M}(T).$$

Here, $\gamma = 2n^{2m+2}$ as in the deformation theorem.

Remarks That T bounds some rectifiable current S is shown by taking the cone over T. The value of the isoperimetric inequality lies in the numerical estimate on $\mathbf{M}(S)$. It was long conjectured that the worst case (exhibiting the best constant) was the sphere, in all dimensions and codimensions. This conjecture was proven in 1986 by Almgren [2]. For merely stationary surfaces, an isoperimetric inequality still holds [Allard, §7.1], but the sharp constant remains conjectural, even for minimal surfaces in $\mathbf{R}^3$. For more general (bounded) integrands than area, an isoperimetric inequality follows trivially for minimizers but remains conjectural for stationary surfaces, even in $\mathbf{R}^3$.

The isoperimetric theorem, 5.3, extends from $\mathbf{R}^n$ to a smooth compact Riemannian manifold $\mathbf{M}$. One must assume that T is a boundary, and the constant γ depends on $\mathbf{M}$. A cycle of small mass is a boundary. These extensions are special cases of the isoperimetric inequality of 12.3, with $\mathbf{B} = \emptyset$.

For more on these results, see Chapter 17 and Chavel [1].

The proof of 5.3 is a bizarre application of the deformation theorem, $T = P + Q + \partial S$. One chooses ε *large*, a grid large enough to force the approximation P to T to be 0. See Figure 5.3.1.

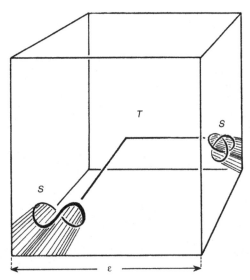

Figure 5.3.1. In the proof of the isoperimetric inequality, T is projected onto a *large* grid.

Proof   Choose ε so that $\gamma \mathbf{M}(T) = \varepsilon^m$, and apply the deformation theorem to obtain $T = P + Q + \partial S$. Since $\partial T = 0$, $Q = 0$. By 5.1(2), $\mathbf{M}(P) \le \gamma \mathbf{M}(T)$. But, since P lies in the 2ε grid, $\mathbf{M}(P)$ must be a multiple of $(2\varepsilon)^m$, which exceeds $\gamma \mathbf{M}(T)$ by choice of ε. Therefore, $P = 0$. Now $T = \partial S$, and, by 5.1(2),

$$\mathbf{M}(S) \le \varepsilon \gamma \mathbf{M}(T) = \varepsilon^{m+1} = [\gamma \mathbf{M}(T)]^{(m+1)/m}. \qquad \blacksquare$$

5.4   The Closure Theorem [Federer, 4.2.16]

(1) $\mathbf{I}_m$ is $\mathbf{F}$ *closed in* $\mathbf{N}_m$,

(2) ("boundary rectifiability") $\mathbf{I}_{m+1} = \{T \in \mathscr{R}_{m+1} : \mathbf{M}(\partial T) < \infty\}$,

(3) $\mathscr{R}_m = \{T \in \mathscr{F}_m : \mathbf{M}(T) < \infty\}$.

Consequently,

(4) $\mathscr{T} = \{T \in \mathbf{I}_m : \operatorname{spt} T \subset \mathbf{B}^n(\mathbf{0}, R), \mathbf{M}(T) \le c, \text{ and } \mathbf{M}(\partial T) \le c\}$ is $\mathscr{F}$ *complete.*

Remarks   This result, specifically (1), is the difficult part of the compactness theorem. It depends on Lemma 4.12, the characterization of rectifiable sets by rectifiable slices. The original proof of Lemma 4.12 used structure theory. However, in 1986 White [3] found a simpler and more direct argument.

It follows directly from the definition of $\mathscr{F}_m$ that $\{T \in \mathscr{F}_m : \operatorname{spt} T \subset \mathbf{B}(\mathbf{0}, R)\}$ is $\mathscr{F}$ complete. Consequence (4) now follows by the lower semicontinuity of $\mathbf{M}$ (Exercise 4.22) and the continuity of ∂.

Proof   We leave it as an exercise (5.2) to check that for each m, $(1) \Rightarrow (2) \Rightarrow (3)$. Hence, to prove (1), (2), and (3), it suffices to prove (1) for m, assuming all three conclusions for $m - 1$.

To prove (1), suppose that a sequence of integral currents $Q_i \in \mathbf{I}_m$ converges in the $\mathbf{F}$ norm to a normal current $T \in \mathbf{N}_m$. We must show that $T \in \mathbf{I}_m$.

By induction, we may assume that $\partial T \in \mathbf{I}_{m-1}$. By replacing T with $T - T_1$, where $T_1 \in \mathbf{I}_m$ has the same boundary as T, we may assume that $\partial T = 0$.

By Lemma 4.12, it suffices to show that for all points $p \in \mathbf{R}^n$, for almost every positive real number r, the slice $\partial (T \lfloor \mathbf{B}(p, r))$ is rectifiable. We may assume that the Q_i's converge so rapidly that

$$\sum \mathbf{F}(Q_i - T) < \infty.$$

Thence by slicing theory 4.11(6), for $0 < a < b$.

$$\int_a^b \sum \mathbf{F}[(Q_i - T) \lfloor \mathbf{B}(p, r)] \, dr < \infty.$$

Therefore, $Q_i \llcorner \mathbf{B}(p, r) \xrightarrow{\mathbf{F}} T \llcorner \mathbf{B}(p, r)$ and hence

$$\partial(Q_i \llcorner \mathbf{B}(p, r)) \xrightarrow{\mathbf{F}} \partial(T \llcorner \mathbf{B}(p, r))$$

for almost every r. Recall that by slicing theory 4.11(5), $\partial(Q_i \llcorner \mathbf{B}(p, r))$ and $\partial(T \llcorner \mathbf{B}(p, r))$, and hence of course $Q_i \llcorner \mathbf{B}(p, r)$ and $T \llcorner \mathbf{B}(p, r)$, are normal currents. By induction on (2), $Q_i \llcorner \mathbf{B}(p, r)$ and hence $\partial(Q_i \llcorner \mathbf{B}(p, r))$ are integral currents. By induction on (1), $\partial(T \llcorner \mathbf{B}(p, r))$ is an integral current (for almost all r). Now by Lemma 4.12, T is a rectifiable current. Since $\partial T = 0$, T is an integral current, and (1) is proved. As mentioned, (2) and (3) follow.

To prove (4), let T_j be a Cauchy sequence in $\mathscr{F}$. By the completeness of $\{T \in \mathscr{F}_m : \operatorname{spt} T \subset \mathbf{B}(\mathbf{0}, R)\}$, there is a limit $T \in \mathscr{F}_m$. By the lower semicontinuity of mass (Exercise 4.22), $\mathbf{M}(T) \leq c$ and $\mathbf{M}(\partial T) \leq c$. Finally, by (3), $T \in \mathbf{I}_m$. ∎

5.5 The Compactness Theorem [Federer, 4.2.17] *Let K be a closed ball in $\mathbf{R}^n$, $0 \leq c < \infty$. Then*

$$\{T \in \mathbf{I}_m \mathbf{R}^n : \operatorname{spt} T \subset K, \ \mathbf{M}(T) \leq c, \ and \ \mathbf{M}(\partial T) \leq c\}$$

is $\mathscr{F}$ compact.

Remark More generally, K may be a compact C^1 submanifold of $\mathbf{R}^n$ or a compact Lipschitz neighborhood retract (cf. Federer [1, 4.1.29]), yielding the compactness theorem in any C^1 compact Riemannian manifold M. (Any C^1 embedding of M in $\mathbf{R}^n$, whether or not isometric, will do, since altering the metric only changes the flat norm by a bounded amount and does not change the topology, as Brian White pointed out to me.)

Proof The set is complete and totally bounded by the closure theorem, 5.4, and the deformation theorem corollary, 5.2. ∎

As an example of the power of the compactness theorem, we prove the following corollary.

5.6 The Existence of Area-Minimizing Surfaces *Let B be an $(m - 1)$-dimensional rectifiable current in $\mathbf{R}^n$ with $\partial B = 0$. Then there is an m-dimensional area-minimizing rectifiable current S with $\partial S = B$.*

Remarks S *area minimizing* means that, for any rectifiable current T with $\partial T = \partial S$, $\mathbf{M}(S) \leq \mathbf{M}(T)$. That B bounds some rectifiable current is shown by taking the cone over B. Even if B is a submanifold, S is not in general.

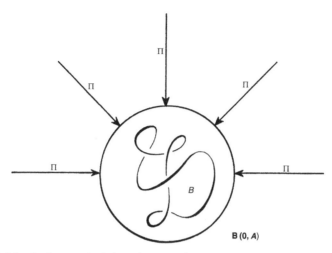

Figure 5.6.1. In the proof of the existence of an area-minimizing surface with given boundary B, it is necessary to keep surfaces under consideration inside some large ball $\mathbf{B}(0, A)$. This is accomplished by projecting everything outside the ball onto its surface.

Proof   Let $\mathbf{B}(0, A)$ be a large ball containing spt B. Let S_j be a sequence of recti-fiable currents with areas decreasing to $\inf\{\mathbf{M}(S): \partial S = B\}$. The first problem is that the spt S_j may send tentacles out to infinity. Let Π denote the Lipschitz map which leaves the ball $\mathbf{B}(0, A)$ fixed and radially projects points outside the ball onto the surface of the ball (Figure 5.6.1). Π is distance nonincreasing and hence area nonincreasing. Therefore, by replacing S_j by $\Pi_{\sharp}S_j$, we may assume that spt $S_j \subset \mathbf{B}(0, A)$.

Now using the compactness theorem we may extract a subsequence which converges to a rectifiable current S. By the continuity of ∂ and the lower semi-continuity of mass (Exercise 4.22), $\partial S = B$ and $\mathbf{M}(S) = \inf\{\mathbf{M}(T): \partial T = B\}$. Therefore, S is the desired area-minimizing surface.   ∎

5.7   The Existence of Absolutely and Homologically Minimizing Surfaces in Manifolds [Federer, 5.1.6]   *Let M be a compact, C^1 Riemannian manifold. Let T be a rectifiable current in M. Then among all recti-fiable currents S in M such that $\partial S = \partial T$ (respectively, $S - T = \partial X$ for some rectifiable current X in M), there is one of least area.*

S is called absolutely or homologically area minimizing. The methods also treat free boundary problems (cf. 12.3).

Proof   Given the *Remark* after the compactness theorem, 5.5, we just need to check that a minimizing limit stays in the same homology class. If $\mathscr{F}(S_i - S)$ is small, $S_i - S = A + \partial B$, with $\mathbf{M}(A)$ and $\mathbf{M}(B)$ small. Let Y_1 be the area minimizer in $\mathbf{R}^n$ with $\partial Y_1 = A$. Since $\mathbf{M}(Y_1)$ is small, by monotonicity 9.5, Y_1 stays close to M and hence may be retracted onto Y in M. Since $S_i - S = \partial Y + \partial B$, therefore $S \sim S_i$, as desired.   ■

Exercises

5.1   Verify that the isoperimetric inequality, 5.3, is homothetically invariant. (A *homothety* μ_r of $\mathbf{R}^n$ maps x to rx.)

5.2   Prove that 5.4(1) $\Rightarrow$ (2) $\Rightarrow$ (3).

5.3   Try to find a counterexample to the closure theorem, 5.4. (Of course, there is none.)

5.4   Show that the isoperimetric inequality, 5.3, fails for normal currents.

Examples of Area-Minimizing Surfaces

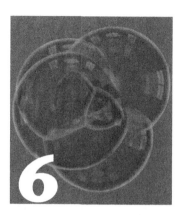

It can be quite difficult to prove that any particular surface is area minimizing. After all, it must compare favorably with all other surfaces with the given boundary. Fortunately, there are a number of beautiful examples. Incidentally, there has been tremendous progress on the classification of various categories of minimal surfaces; see, for example, Meeks and Pérez.

6.1 The Minimal Surface Equation [Federer, 5.4.18] *Let f be a C^2, real-valued function on a planar domain D, such that the graph of f is area minimizing. Then f satisfies the minimal surface equation:*

$$(1 + f_y^2) f_{xx} - 2 f_x f_y f_{xy} + (1 + f_x^2) f_{yy} = 0.$$

Conversely, if f satisfies the minimal surface equation on a convex domain, then its graph is area minimizing.

The minimal surface equation just gives the necessary condition that under smooth variations in the surface, the rate of change of the area is 0. This condition turns out to be equivalent to the vanishing of the mean curvature. A smoothly immersed surface which is locally the graph of a solution to the minimal surface equation (or, equivalently, which has mean curvature 0) is called a *minimal surface*. Some famous minimal surfaces are shown in Figures 6.1.1–6.1.3b.

Theorem 6.1 guarantees that small pieces of minimal surfaces are area minimizing, but larger pieces may not be. For example, the portion of Enneper's surface shown in Figure 6.1.2 is not area minimizing. There are two area-minimizing surfaces with the same boundary, shown in Figure 6.1.4. Some systems of curves in $\mathbf{R}^3$ bound infinitely many minimal surfaces. See Morgan [6] and references therein or the popular articles by Morgan [18, 21].

On the disc (or any other convex domain), there is a solution of the minimal surface equation with any given continuous boundary values. We omit the proof.

Geometric Measure Theory. http://dx.doi.org/10.1016/B978-0-12-804489-6.00006-X

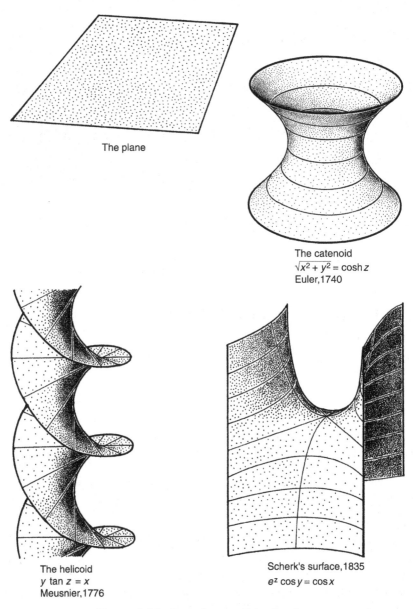

The plane

The catenoid
$\sqrt{x^2 + y^2} = \cosh z$
Euler,1740

The helicoid
$y \tan z = x$
Meusnier,1776

Scherk's surface,1835
$e^z \cos y = \cos x$

Figure 6.1.1. Some famous minimal surfaces.

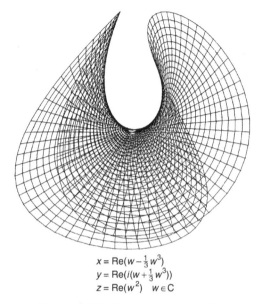

$$x = \mathrm{Re}(w - \tfrac{1}{3}w^3)$$
$$y = \mathrm{Re}(i(w + \tfrac{1}{3}w^3))$$
$$z = \mathrm{Re}(w^2) \quad w \in \mathbf{C}$$

Figure 6.1.2. Enneper's surface, 1864.

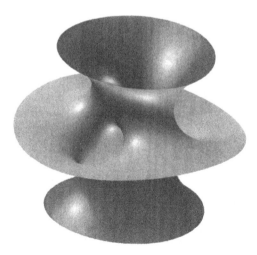

Figure 6.1.3a. The first modern complete, embedded minimal surface of Costa and Hoffman and Meeks (also see Hoffman). Courtesy of David Hoffman, Jim Hoffman, and Michael Callahan.

Figure 6.1.3b. One of the latest new complete, embedded minimal surfaces: the genus one helicoid, discovered by David Hoffman, Hermann Karcher, and Fusheng Wei (1993). Proved embedded by Weber, Hoffman, and Wolf. Computer-generated image by James T. Hoffman at the GANG Laboratory, University of Massachusetts, Amherst. Copyright GANG, 1993.

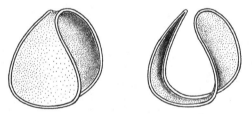

Figure 6.1.4. Area-minimizing surfaces with the same boundary as Enneper's surface.

Because the minimal surface equation is not linear, this fact is not at all obvious, and it fails if the domain is not convex. Moreover, on nonconvex domains there can be solutions of the minimal surface equation whose graphs are not area minimizing (Figure 6.1.5).

The first part of the proof of 6.1 will afford an opportunity to illustrate the classical method and notation of the calculus of variations. In general, begin with a function $f : D \to \mathbf{R}^m$, supposed to maximize or minimize some functional $A(f)$ for the given boundary values $f|\partial D$. Consider infinitesimal changes δf in f, with $\delta f|\partial D = 0$. Set to 0 the corresponding change in A, called the first variation δA. Use integration by parts to obtain an equation of the form

$$\int_D G(f) \cdot \delta f = 0.$$

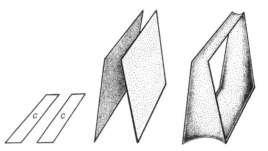

Figure 6.1.5. A minimal graph over a *nonconvex* region C need not be area minimizing. The second surface has less area.

Since this equation holds for all admissible δf, it follows immediately that $G(f) = 0$. This turns out to be a differential equation for f, called the Euler–Lagrange equation. When $A(f)$ is the area of the graph of f, the associated Euler–Lagrange equation is the minimal surface equation.

Proof of 6.1 First, we will show that if the graph of f is area minimizing, then f satisfies the minimal surface equation. By hypothesis, f minimizes the area functional

$$A(f) = \int_D (1 + f_x^2 + f_y^2)^{1/2} \, dx \, dy$$

for given boundary values. Therefore, the variation δA in A due to an infinitesimal, smooth variation δf in f, with $\delta f | \partial D = 0$, must vanish:

$$0 = \delta A = \int_D \frac{1}{2}(1 + f_x^2 + f_y^2)^{-1/2}(2 f_x \delta f_x + 2 f_y \delta f_y) \, dx \, dy$$

$$= \int_D ([(1 + f_x^2 + f_y^2)^{-1/2} f_x] \delta f_x + [(1 + f_x^2 + f_y^2)^{-1/2} f_y] \delta f_y) \, dx \, dy.$$

Integration by parts now yields an equation of the form $\int_D G(f) \delta f = 0$, where

$$G(f) = -\frac{\partial}{\partial x}[(1 + f_x^2 + f_y^2)^{-1/2} f_x] - \frac{\partial}{\partial y}[(1 + f_x^2 + f_y^2)^{-1/2} f_y]$$

$$= \frac{1}{2}(1 + f_x^2 + f_y^2)^{-3/2}(2 f_x f_{xx} + 2 f_y f_{xy}) f_x - (1 + f_x^2 + f_y^2)^{-1/2} f_{xx}$$

$$+ \frac{1}{2}(1 + f_x^2 + f_y^2)^{-3/2}(2 f_x f_{xy} + 2 f_y f_{yy}) f_y - (1 + f_x^2 + f_y^2)^{-1/2} f_{yy}$$

$$= -(1 + f_x^2 + f_y^2)^{-3/2}[(1 + f_y^2) f_{xx} - 2 f_x f_y f_{xy} + (1 + f_x^2) f_{yy}].$$

Since $\int_D G(f)\delta f = 0$ for all smooth δf satisfying $\delta f | \partial D = 0$, $G(f) = 0$; that is,

$$(1 + f_y^2)f_{xx} - 2f_x f_y f_{xy} + (1 + f_x^2)f_{yy} = 0,$$

the minimal surface equation.

Second, we prove that the graph over a convex domain of a solution to the minimal surface equation is area minimizing. This proof uses a *calibration*—that is, a differential form φ which is closed ($d\varphi = 0$) and has maximum value sup $\varphi(\xi) = 1$ as a function on the set of unit k-planes ξ. A surface is said to be calibrated by φ if each oriented tangent plane ξ satisfies $\varphi(\xi) = 1$. The method shows that a calibrated surface is automatically area minimizing.

Given $f : D \to \mathbf{R}$, define a 2-form φ on $D \times \mathbf{R}$ by

$$\varphi(x, y) = \frac{-f_x dy\, dz - f_y\, dz\, dx + dx\, dy}{\sqrt{f_x^2 + f_y^2 + 1}}.$$

Then for any point $(x, y, z) \in D \times \mathbf{R}$ and for any unit 2-vector ξ, $\varphi(\xi) \leq 1$, with equality if ξ is tangent to the graph of f at $(x, y, f(x, y))$. Moreover, φ is closed:

$$d\varphi = -\frac{\partial}{\partial x} f_x (f_x^2 + f_y^2 + 1)^{-1/2} - \frac{\partial}{\partial y} f_y (f_x^2 + f_y^2 + 1)^{-1/2}\, dx\, dy\, dz$$

$$= -(f_x^2 + f_y^2 + 1)^{-3/2}((1 + f_y^2)f_{xx} - 2f_x f_y f_{xy} + (1 + f_x^2)f_{yy})\, dx\, dy\, dz$$

$$= 0$$

by the minimal surface equation.

Now let S denote the graph of f, and let T be any other rectifiable current with the same boundary. Since D is convex, we may assume spt $T \subset D \times \mathbf{R} =$ domain φ, by projecting T into $D \times \mathbf{R}$ if necessary without increasing area T. Now, since $\varphi(\xi) = 1$ whenever ξ is tangent to S,

$$\text{area } S = \int_S \varphi.$$

Since $S - T$ bounds and φ is closed,

$$\int_S \varphi = \int_T \varphi.$$

Since $\varphi(\xi) \leq 1$ for all 2-planes ξ,

$$\int_T \varphi \leq \text{area } T.$$

Combining the two equations and the inequality yields

$$\text{area } S \leq \text{area } T.$$

Therefore, S is area minimizing, as desired.   ∎

Remark   The same argument with the same calibration shows that if the graph of f has constant mean curvature, then it minimizes area for fixed volume constraint in the cylinder over the domain.

6.2   Remarks on Higher Dimensions   For a function $f : \mathbf{R}^{n-1} \to \mathbf{R}$, the minimal surface equation takes the form

$$(1) \qquad \qquad \operatorname{div} \frac{\nabla f}{\sqrt{1 + |\nabla f|^2}} = 0.$$

The statement and proof of 6.1 apply virtually unchanged.

In higher codimension there is a minimal surface system. For example, for a function $f : \mathbf{R}^2 \to \mathbf{R}^{n-2}$, the minimal surface system is

$$(2) \qquad (1 + |f_y|^2) f_{xx} - 2(f_x \cdot f_y) f_{xy} + (1|f_x|^2) f_{yy} = 0.$$

In general codimension, the graph of f need not be area minimizing, even if the domain is convex. (In an attempted generalization of the proof of 6.1, φ generally would not be closed. For a counterexample, see Lawson and Osserman).

6.3   Complex Analytic Varieties [Federer, 5.4.19]   *Any compact portion of a complex analytic variety in $\mathbf{C}^n \cong \mathbf{R}^{2n}$ is area minimizing.*

With this initially astonishing fact, Federer provided some of the first examples of area-minimizing surfaces with singularities. For example, $\{w^2 = z^3\} \subset \mathbf{C}^2$ has an isolated "branch point" singularity at 0. Similarly, the union of the complex axis planes $\{z = 0\}$ and $\{w = 0\}$ in $\mathbf{C}^2$ has an isolated singularity at 0. These are probably the simplest examples. $\{w = \pm\sqrt{z}\}$ is regular at 0; it is the graph of $z = w^2$. The proof, as that of the second part of 6.1, uses a "calibration" φ.

Proof   On $\mathbf{C}^n = \mathbf{R}^n \oplus \mathbf{R}^n$, let ω be the Kähler form

$$\omega = dx_1 \wedge dy_1 + \cdots + dx_n \wedge dy_n.$$

Wirtinger's inequality [Federer, 1.8.2, p. 40] states that the real $2p$-form $\varphi = \omega^p / p!$ satisfies

$$|\varphi(\xi)| \leq 1$$

for every $2p$-plane ξ, with equality if and only if ξ is a complex p-plane.

Now let S be a compact portion of a p-dimensional complex analytic variety, and let T be any other $2p$-dimensional rectifiable current with the same boundary. Since S is complex analytic, $\varphi(\xi)$ is 1 on every plane tangent to S, and

$$\text{area } S = \int_S \varphi.$$

Since $\partial S = \partial T$, $S - T$ is a boundary. Of course, $d\varphi = 0$ because φ has constant coefficients. Therefore,

$$\int_S \varphi = \int_T \varphi.$$

Finally, since $\varphi(\xi)$ is always at most 1,

$$\int_T \varphi \le \text{area } T.$$

Combining the equalities and inequality yields

$$\text{area } S \le \text{area } T.$$

We conclude that S is area minimizing, as desired.

The second part of the proof of 6.1 and 6.3 proved area minimization by means of a calibration or closed differential form of comass 1, as examples of the following method. ∎

6.4 Fundamental Theorem of Calibrations   *Let φ be a closed differential form of unit comass in $\mathbf{R}^n$ or in any smooth Riemannian manifold M. Let S be an integral current such that $\langle \vec{S}, \varphi \rangle = 1$ at almost all points of S. In $\mathbf{R}^n$, S is area minimizing for its boundary. In any M, S is area minimizing in its homology class (with or without boundary).*

Proof   Let T be any comparison surface. Then

$$\text{area } S = \int_S \varphi = \int_T \varphi \le \text{area } T. \qquad ∎$$

6.5 History of Calibrations (cf. Morgan [1, 2])   The original example of complex analytic varieties was implicit in Wirtinger (1936), explicit for complex analytic submanifolds in de Rham [1] (1957), and applied to singular complex varieties in the context of rectifiable currents by Federer [3, §4] (1965).

Berger [2, §6, last paragraph] (1970) was the first to extract the underlying principle and apply it to other examples such as quaternionic varieties, followed by Dao (1977). The term *calibration* was coined in the landmark paper of Harvey and Lawson, which discovered rich new calibrated geometries of "special Lagrangian," "associative," and "Cayley" varieties.

The method has grown in power and applications. Surveys appear in Morgan [1, 2], Harvey, and Joyce. Mackenzie and Lawlor use calibrations in the proof (Nance; Lawlor [1]) of the angle conjecture on when a pair of m-planes in $\mathbf{R}^n$ is area minimizing. The "vanishing calibrations" of Lawlor [3] actually provide sufficient differential–geometric conditions for area minimization, a classification of all area-minimizing cones over products of m spheres, examples of nonorientable area-minimizing cones, and singularities stable under perturbations. The "paired calibrations" of Lawlor and Morgan [2] and of Brakke [1, 2] and the covering space calibrations of Brakke [3] prove new examples of soap films in $\mathbf{R}^3$, in $\mathbf{R}^4$, and above. Other developments include Murdoch's "twisted calibrations" of nonorientable surfaces, Le's "relative calibrations" of stable surfaces, and Pontryagin calibrations on Grassmannians (Gluck, Mackenzie, and Morgan).

Lawlor [2] has developed a related theory for proving minimization by slicing. Lawlor and Morgan [1] show, for example, that three minimal surfaces meeting at 120 degrees minimize area locally.

Exercises

6.1 Verify that the helicoid is a minimal surface.

6.2 Verify that Scherk's surface is a minimal surface.

6.3 Verify that Enneper's surface is a minimal surface.

6.4 Prove that the catenoids $\sqrt{x^2 + y^2} = \frac{1}{a} \cosh az$ are the only smooth, minimal surfaces of revolution in $\mathbf{R}^3$.

6.5 Verify that for $n = 3$ the minimal surface equation, 6.2(1), reduces to that of 6.1.

6.6 Use 6.2(2) to verify that the complex analytic variety

$$\{w = z^2\} \subset \mathbf{C}^2$$

is a minimal surface.

6.7 Use 6.2(2) to prove that the graph of any complex analytic function $g: \mathbf{C} \to \mathbf{C}$ is a minimal surface.

The Approximation Theorem

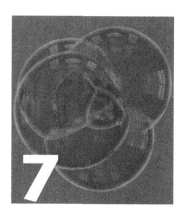

The approximation theorem states that an integral current, T, can be approximated by a slight diffeomorphism of a polyhedral chain, P, or, equivalently, that a slight diffeomorphism $f_\sharp T$ of T can be approximated by P itself. (See Figure 7.1.1.) The approximation P actually coincides with $f_\sharp T$ except for an error term, E, of small mass (whereas the deformation theorem, 5.1, provided an approximation only in the flat norm).

7.1 The Approximation Theorem [Federer, 4.2.20] *Given an integral current $T \in \mathbf{I}_m \mathbf{R}^n$ and $\varepsilon > 0$, there exist a polyhedral chain $P \in \mathscr{P}_m \mathbf{R}^n$, supported within a distance ε of the support of T, and a C^1 diffeomorphism f of $\mathbf{R}^n$ such that*

$$f_\sharp T = P + E$$

with $\mathbf{M}(E) \leq \varepsilon$, $\mathbf{M}(\partial E) \leq \varepsilon$, $\mathrm{Lip}(f) \leq 1+\varepsilon$, $\mathrm{Lip}(f^{-1}) \leq 1+\varepsilon$, $|f(x)-x| \leq \varepsilon$, *and* $f(x) = x$ *whenever* $\mathrm{dist}(x, \mathrm{spt}\, T) \geq \varepsilon$.

Proof. **CASE 1** *∂T polyhedral.* Since T is rectifiable, $T = (\mathscr{H}^m \llcorner B) \wedge \zeta$, with B rectifiable and $|\zeta|$ integer valued. By Proposition 3.11, $\mathscr{H}^m$ almost all of B is contained in a countable union $\cup\, M_i$ of disjoint C^1 embedded manifolds. At almost every point $x \in B$, the density of B and of $\cup\, M_i$ is 1 (Proposition 3.12) so that there is a single M_i such that B and M_i coincide at x except for a set of density 0. Now a covering argument produces a finite collection of disjoint open balls $U_i \subset \mathbf{R}^n - \mathrm{spt}\,\partial T$ and nearly flat C^1 submanifolds N_i of U_i such that $\cup\, N_i$ coincides with B except for a set of small $\|T\|$ measure. Gentle hammering inside each U_i, smoothed at the edges, is the desired diffeomorphism f of $\mathbf{R}^n$, which flattens most of B into m-dimensional planes, where $f_\sharp T$ can be $\mathbf{M}$ approximated by a polyhedral chain P_1 (cf. proof of 4.4).

Geometric Measure Theory. http://dx.doi.org/10.1016/B978-0-12-804489-6.00007-1

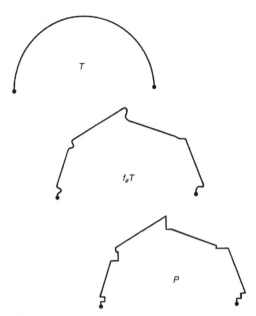

Figure 7.1.1. The approximation theorem yields a diffeomorphism $f_\sharp T$ of T which coincides with a polyhedral P except for small measure.

Unfortunately, the error $f_\sharp T - P_1$, although small in mass, may have huge boundary. Now we use the deformation theorem, 5.1, to decompose the error

$$f_\sharp T - P_1 = P_2 + Q + \partial S.$$

As usual, Q and S have small mass. But in this case, because $f_\sharp T - P_1$ has small mass, so does P_2 and hence so does the remaining term, ∂S. Because f leaves spt ∂T fixed, $\partial(f_\sharp T - P_1) = \partial T - \partial P_1$ is polyhedral, and hence the deformation theorem construction makes Q polyhedral. Now take

$$P = P_1 + P_2 + Q.$$

Then $f_\sharp T - P = \partial S$ has small mass and no boundary.

GENERAL CASE When ∂T is not polyhedral, first approximate ∂T by the first case,

$$f_{1\sharp} \partial T = P_1 + \partial S_1,$$

with $\mathbf{M}(S_1)$ and $\mathbf{M}(\partial S_1)$ small. Now $f_{1\sharp}T - S_1$ has polyhedral boundary and can be approximated by an integral polyhedral chain, P_2,

$$f_{2\sharp}(f_{1\sharp}T - S_1) = P_2 + \partial S_2,$$

with $\mathbf{M}(\partial S_2)$ small. Therefore,

$$(f_2 \circ f_1)_{\sharp}T = P_2 + (f_{2\sharp}S_1 + \partial S_2),$$

and the error term and its boundary have small mass, as desired.   ∎

Survey of Regularity Results

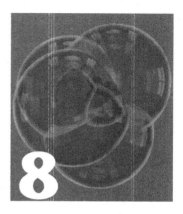

In 1962, Wendell Fleming proved a regularity result that at first sounds too good to be true.

8.1 Theorem [Fleming 2] *A two-dimensional, area-minimizing rectifiable current T in $\mathbf{R}^3$ is a smooth, embedded manifold on the interior.*

More precisely, spt $T -$ spt ∂T is a C^∞ embedded manifold.

In the classical theory, such complete regularity fails. The disc of least mapping area with given boundary is an immersed minimal surface (see Nitsche [2, §365, p. 318], Osserman [2, Appendix 3, §1, p. 143], or Lawson [§3, p. 76]) but not in general embedded. For the boundary shown in Figure 8.1.1, a circle with a tail, the area-minimizing rectifiable current has higher genus, has less area, and is embedded. Pictured in Figure 8.1.2, it flows from the top, flows down the tail, pans out in back onto the disc, flows around front, and flows down the tail to the bottom. There is a hole in the middle that you can stick your finger through. Incidentally, this surface exists as a soap film, whereas the least-area disc does not.

The regularity theorem 8.1 was generalized to three-dimensional surfaces in $\mathbf{R}^4$ by Almgren [6] in 1966 and up through six-dimensional surfaces in $\mathbf{R}^7$ by Simons in 1968. In 1969, Bombieri, De Giorgi, and Giusti gave an example of a seven-dimensional, area-minimizing rectifiable current in $\mathbf{R}^8$ with an isolated interior singularity. Chapter 10 gives a short discussion of this counterexample as well as an outline of the proof of the regularity results.

The complete interior regularity results for area-minimizing hypersurfaces are given by the following theorem of Federer.

8.2 Theorem [Federer 2] *An $(n-1)$-dimensional, area-minimizing rectifiable current T in $\mathbf{R}^n$ is a smooth, embedded manifold on the interior except for*

Geometric Measure Theory. http://dx.doi.org/10.1016/B978-0-12-804489-6.00008-3

Figure 8.1.1. A least-area disc need not be embedded.

Figure 8.1.2. The area-minimizing rectifiable current is embedded.

a singular set of Hausdorff dimension at most $n - 8$; if $n = 8$, singularities are isolated points.

Regularity in higher codimension, for an m-dimensional area-minimizing rectifiable current T in $\mathbf{R}^n$, with $m < n - 1$, is much harder. Until 1983 it was known only that the set of regular points, where spt T is a smooth embedded manifold, was dense in spt $T -$ spt ∂T [Federer, 5.3.16]. On the other hand, m-dimensional complex analytic varieties, which are automatically area minimizing (6.3), can have $(m - 2)$-dimensional singular sets. In a major advance, Almgren proved the conclusive regularity theorem.

8.3   Theorem [Almgren 3, 1983]   *An m-dimensional, area-minimizing rectifiable current in $\mathbf{R}^n$ is a smooth, embedded manifold on the interior except for a singular set of Hausdorff dimension at most $m - 2$.*

For example, a two-dimensional area-minimizing rectifiable current in $\mathbf{R}^n$ has at worst a zero-dimensional interior singular set. In 1988, Sheldon Chang (Figure 8.3.1) proved that these singularities must be isolated, "classical branch points."

The stronger regularity theory in codimension 1 comes from an elementary reduction to the relatively easy case of surfaces of multiplicity 1. Indeed, a nesting lemma decomposes an area-minimizing hypersurface into nested, multiplicity-1

Figure 8.3.1. Graduate student Sheldon Chang sharpened Fred Almgren's regularity results for two-dimensional area-minimizing surfaces. Pictured outside the Almgren house in Princeton. Photo courtesy Jean Taylor.

area-minimizing surfaces, for which strong regularity results hold. If these surfaces touch, they must coincide by a maximum principle. (See Section 8.5 and Chapter 10.)

In general dimensions and codimensions, very little is known about the structure of the set S of singularities. One might hope that S stratifies into embedded manifolds of various dimensions. However, for all we know to date, S could even be fractional dimensional.

8.4 Boundary Regularity   In 1979, Hardt and Simon proved the conclusive boundary regularity theorem for area-minimizing hypersurfaces.

Theorem [Hardt and Simon 1]   *Let T be an $(n-1)$-dimensional, area-minimizing rectifiable current in $\mathbf{R}^n$, bounded by a C^2, oriented submanifold (with multiplicity 1). Then at every boundary point, spt T is a C^1, embedded manifold-with-boundary.*

Notice that general regularity is stronger at the boundary than on the interior, where there can be an $(n - 8)$-dimensional singular set. The conclusion is meant to include the possibility that spt T is an embedded manifold without boundary at some boundary points, as occurs, for example, if the given boundary is two concentric, similarly oriented planar circles. (See Figure 8.4.1.)

The surface of Figure 8.1.2 at first glance may seem to have a boundary singularity where the tail passes through the disc. Actually, as you move down the tail, the inward conormal rotates smoothly and rapidly almost a full 360°.

B. White [5] has generalized regularity to smooth boundaries with multiplicities.

In higher codimension, boundary singularities can occur. For example, in $\mathbf{C}^2$, the union of a half unit disc in the z-plane and a unit disc in the w-plane is area minimizing (6.3) and has a boundary singularity at the origin.

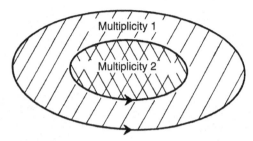

Figure 8.4.1. This example shows that part of the given boundary can end up on the interior of the area-minimizing surface T.

For the classical least-area disc in $\mathbf{R}^3$, it remains an open question whether there can be boundary branch points if the boundary is C^∞ but not real analytic (see Nitsche [2, §366, p. 320]).

8.5 General Ambients, Volume Constraints, and Other Integrands
All of the previous regularity results continue to hold in smooth Riemannian manifolds other than $\mathbf{R}^n$. Federer [1, 5.4.15, 5.4.4] gives the proof of Theorem 8.2 for $\mathbf{R}^n$ and explains that the methods can be generalized to any smooth Riemannian manifold, except possibly a maximum principle. Because two regular minimal surfaces in local coordinates are graphs of functions satisfying a certain nonlinear, elliptic, partial differential system (see Morgan [7, §2.2] or Federer [1, 5.1.11]), such a maximum principle is standard, due essentially to Hopf [Satz 1'; see also Serrin, p. 184].

Regularity also holds for a hypersurface minimizing area subject to a volume constraint because the tangent cone is area minimizing without volume constraint. Federer [2, Lemma 2] shows that the Hausdorff dimension of the singular set is less than or equal to the Hausdorff dimension of the singular set of some area-minimizing tangent cone. It is enough for the metric to be Lipschitz [Morgan 27]. All of these results hold as well in manifolds with density [Morgan 27, §3.10]; see Chapter 18.

Incidentally, Schoen, Simon, and Yau, as well as Schoen and Simon, have proved regularity results for hypersurfaces that are stable but not necessarily area minimizing in Riemannian manifolds of dimension less than eight.

Regularity for surfaces minimizing more general integrands than area is harder, and the results are weaker. For a smooth elliptic integrand Φ (see 12.5) on an n-dimensional smooth Riemannian manifold, an m-dimensional Φ-minimizing rectifiable current is a smooth embedded manifold on an open, dense set [Federer 1, 5.3.17]. If $m = n - 1$, the interior singular set has dimension less than $m - 2$ (Schoen, Simon, and Almgren [II.7, II.9] plus an additional unpublished argument of Almgren as in White [7, Theorem 5.2]). The achievement of such general regularity results is one of the telling strengths of geometric measure theory.

Exercises

8.1 Try to come up with a counterexample to Theorems 8.1 and 8.4.

8.2 Try to draw an area-minimizing rectifiable current bounded by the trefoil knot. (Make sure your surface is orientable.)

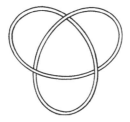

8.3 Illustrate the sharpness of Theorem 8.3 by proving directly that the union of two orthogonal unit discs about 0 in $\mathbf{R}^4$ is area minimizing.

Hint: First show that the area of any surface S is at least the sum of the areas of its projections into the x_1-x_2 and x_3-x_4 planes.

Monotonicity and Oriented Tangent Cones

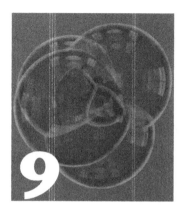

This chapter introduces the two basic tools of the regularity theory of area-minimizing surfaces. Section 9.2 presents the monotonicity of the mass ratio, a lower bound on area growth. Section 9.3 presents the existence of an oriented tangent cone at *every* interior point of an area-minimizing surface.

9.1 Locally Integral Flat Chains [Federer, 4.1.24, 4.3.16]
We need to generalize our definitions to include noncompact surfaces such as oriented planes. First, define the space $\mathscr{F}_m^{\mathrm{loc}}$ of locally integral flat chains as currents which locally coincide with integral flat chains:

$$\mathscr{F}_m^{\mathrm{loc}} = \{T \in \mathscr{D}_m : \text{for all } x \in \mathbf{R}^n \text{ there exists } S \in \mathscr{F}_m \text{ with } x \notin \mathrm{spt}(T - S)\}.$$

For the local flat topology, a typical neighborhood of **0** takes the form

$$U_{\delta,R} = \{T \in \mathscr{F}_m^{\mathrm{loc}} : \mathrm{spt}(T - (A + \partial B)) \cap \mathbf{U}(\mathbf{0},\, R) = \varnothing,$$
$$A \in \mathscr{R}_m,\ B \in \mathscr{R}_{m+1},\ \mathbf{M}(A) + \mathbf{M}(B) < \delta\},$$

where $\mathbf{U}(\mathbf{0}, R)$ is the open ball $\{x \in \mathbf{R}^n : |x| < R\}$.

The subspaces $\mathbf{I}_m^{\mathrm{loc}} \subset \mathscr{R}_m^{\mathrm{loc}} \subset \mathscr{F}_m^{\mathrm{loc}}$ of locally integral currents and locally rectifiable currents are defined analogously:

$$\mathbf{I}_m^{\mathrm{loc}} = \{T \in \mathscr{D}_m : \text{for all } x \in \mathbf{R}^n \text{ there exists } S \in \mathbf{I}_m \text{ with } x \notin \mathrm{spt}(T - S)\},$$
$$\mathscr{R}_m^{\mathrm{loc}} = \{T \in \mathscr{D}_m : \text{for all } x \in \mathbf{R}^n \text{ there exists } S \in \mathscr{R}_m \text{ with } x \notin \mathrm{spt}(T - S)\}.$$

Alternative definitions of the locally rectifiable currents are given by

$$\mathscr{R}_m^{\mathrm{loc}} = \{T \in \mathscr{D}_m : T \llcorner \mathbf{B}(\mathbf{0}, R) \in \mathscr{R}_m \text{ for all } R\}$$
$$= \{T \in \mathscr{D}_m : T \llcorner \mathbf{B}(a, R) \in \mathscr{R}_m \text{ for all } a \text{ and all } \mathbf{R}\}.$$

Geometric Measure Theory. http://dx.doi.org/10.1016/B978-0-12-804489-6.00009-5

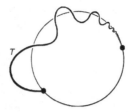

Figure 9.1.1. T is a single curve with two endpoints. However, its restriction to the inside of the circle has infinitely many pieces and infinitely many endpoints. Hence, the restriction of an integral current need not be an integral current.

There are no similar alternatives for locally integral currents. Indeed, Figure 9.1.1 shows an integral current T such that $T \, \llcorner \, \mathbf{B}(\mathbf{0}, 1)$ is not an integral current, because $\mathbf{M}(\partial(T \, \llcorner \, \mathbf{B}(\mathbf{0}, 1))) = +\infty$.

A locally rectifiable current T is called *area minimizing* if for all a and R, $T \, \llcorner \, \mathbf{B}(a, R)$ is area minimizing.

The compactness theorem, 5.5, generalizes to unbounded locally integral currents in $\mathbf{R}^n$ or noncompact manifolds, as explained in Simon [3, 27.3, 31.2]. Without this perspective, Federer [1] must be forever judiciously restricting currents, a troublesome complication.

9.2   Monotonicity of the Mass Ratio   For a locally rectifiable current $T \in \mathscr{R}_m^{\text{loc}}$ and a point $a \in \mathbf{R}^n$, define the mass ratio

$$\Theta^m(T, \, a, \, r) = \mathbf{M}(T \, \llcorner \, \mathbf{B}(a, \, r))/\alpha_m r^m,$$

where α_m is the measure of the unit ball in $\mathbf{R}^m$. Define the density of T at a,

$$\Theta^m(T, \, a) = \lim_{r \to 0} \Theta^m(T, \, a, \, r).$$

The following theorem on the monotonicity of the mass ratio is one of the most useful tools in regularity theory.

9.3   Theorem [Federer, 5.4.3]   *Let T be an area-minimizing locally rectifiable current in $\mathscr{R}_m^{\text{loc}}$. Let a lie in the support of T. Then for $0 < r < \text{dist}(a, \text{spt } \partial T)$, the mass ratio $\Theta(T, \, a, \, r)$ is a monotonically increasing function of r.*

Remark   If the mass ratio is constant, then T is a cone (invariant under homothetic expansions).

Monotonicity for Minimal Surfaces and Other Integrands   Monotonicity actually holds for stationary (minimal) surfaces or, in a weakened form, even for bounded mean curvature surfaces in manifolds, but not for more general integrands than area [Allard, 5.1; see Sect. 11.2]. (Nevertheless, 9.5 holds with a smaller constant for minimizers of general integrands.)

Before giving the proof of 9.3, we state two immediate corollaries.

9.4 Corollary *Suppose $T \in \mathscr{R}_m^{\text{loc}}$ is area minimizing. Then $\Theta^m(T, a)$ exists for every $a \in \text{spt}\, T - \text{spt}\, \partial T$.*

9.5 Corollary *Suppose $T \in \mathscr{R}_m^{\text{loc}}$ is area minimizing and $a \in \text{spt}\, T$. Then for $0 < r < \text{dist}(a, \text{spt}\, \partial T)$,*

$$\mathbf{M}(T \llcorner \mathbf{B}(a, r)) \geq \Theta^m(T, a) \cdot \alpha_m r^m.$$

For example, if furthermore T happens to be an embedded, two-dimensional, oriented manifold-with-boundary, then

$$\mathbf{M}(T \llcorner \mathbf{B}(a, r)) \geq \pi r^2.$$

See Figures 9.5.1 and 9.5.2.

Proof of Theorem 9.3 For $0 < r < \text{dist}(a, \text{spt}\, \partial T)$, let $f(r)$ denote $\mathbf{M}(T \llcorner \mathbf{B}(a, r))$. Since f is monotonically increasing, for almost all r, $f'(r)$ exists. Slicing by the function $u(x) = |x - a|$ yields (4.11(3))

$$\mathbf{M}(\partial(T \llcorner \mathbf{B}(a, r))) \leq f'(r).$$

Since T is area minimizing, $\mathbf{M}(T \llcorner \mathbf{B}(a, r))$ is less than or equal to the area of the cone C over $\partial(T \llcorner \mathbf{B}(a, r))$ to a (Figure 9.5.3).

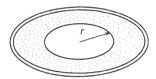

Figure 9.5.1. The disc D is area minimizing and $\mathbf{M}(D \llcorner \mathbf{B}(0, r)) = \pi r^2$.

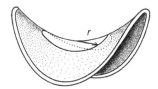

Figure 9.5.2. An area-minimizing surface with $\mathbf{M}(T \llcorner \mathbf{B}(0, r)) > \pi r^2$.

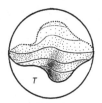

Figure 9.5.3. If $T \llcorner \mathbf{B}(a, r)$ is area minimizing, then it must of course have less area than the cone C over its boundary.

By Exercise 3.8, $\mathbf{M}(C) = \frac{r}{m}\mathbf{M}(\partial(T \llcorner \mathbf{B}(a, r)))$. Assembling these inequalities yields

$$f(r) \leq \mathbf{M}(C) = \frac{r}{m}\mathbf{M}(\partial(T \llcorner \mathbf{B}(a, r))) \leq \frac{r}{m}f'(r).$$

Consequently,

$$\frac{d}{dr}\alpha_m \Theta(T, a, r) = \frac{d}{dr}[r^{-m}f(r)] = r^{-m}f'(r) - mr^{-m-1}f(r)$$

$$= \frac{m}{r^{m-1}}\left[\frac{r}{m}f'(r) - f(r)\right] \geq 0.$$

Hence, the absolutely continuous part of $\Theta(T, a, r)$ is increasing. Since any singular part is due to increases in f, $\Theta(T, a, r)$ is increasing as desired. ∎

9.6 Corollary *Let T be an area-minimizing rectifiable current in $\mathscr{R}_m \mathbf{R}^n$. Then for all $a \in \operatorname{spt} T - \operatorname{spt} \partial T$, $\Theta^m(T, a) \geq 1$.*

Proof Since a rectifiable set has density 1 almost everywhere (3.12), there is a sequence of points $a_j \to a$ with $\Theta^m(T, a_j) \geq 1$. Let $0 < r < \operatorname{dist}(a, \operatorname{spt} \partial T)$, $r_j = \operatorname{dist}(a, a_j)$. Obviously

$$\mathbf{M}(T \llcorner \mathbf{B}(a, r)) \geq \mathbf{M}(T \llcorner \mathbf{B}(a_j, r - r_j)).$$

But by monotonicity, $\mathbf{M}(T \llcorner \mathbf{B}(a_j, r - r_j)) \geq \alpha_m(r - r_j)^m$. Consequently, $\mathbf{M}(T \llcorner \mathbf{B}(a, r)) \geq \alpha_m r^m$ and $\Theta(T, a) \geq 1$. ∎

9.7 Oriented Tangent Cones [Federer, 4.3.16] We now develop a generalization to locally integral flat chains of the notion of the tangent plane to a C^1 manifold at a point. (See Figure 9.7.1.)

Definitions A locally integral flat chain C is called a *cone* if every homothetic expansion or contraction $\mu_{R\sharp}C = C$. If $T \in \mathscr{F}_m^{loc}$, such a cone C is called an *oriented tangent cone to T at $\mathbf{0}$* if there is a decreasing sequence $r_1 > r_2 > r_3 > \cdots$

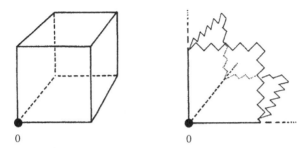

Figure 9.7.1. The surface of the unit cube and the three quarter planes constituting its oriented tangent cone at **0**.

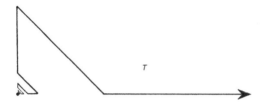

Figure 9.7.2. T alternates ad infinitum between the positive x-axis and the positive y-axis. Each axis is an oriented tangent cone.

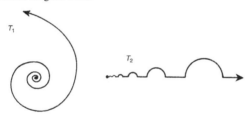

Figure 9.7.3. T_1 and T_2 are both invariant under certain sequences of homothetic expansions but are not cones.

tending to 0 such that $\mu_{r_j^{-1}\sharp}T$ converges to C in the local flat topology. Note that an oriented tangent cone C is a current, whereas a tangent cone $\mathrm{Tan}(E, \mathbf{0})$ as defined in Section 3.9 is a set. In general, spt $C \subset \mathrm{Tan}(\mathrm{spt}\ T, \mathbf{0})$, but equality need not hold (cf. Exercise 9.6).

Remarks   Figure 9.7.2 illustrates that an oriented tangent cone is not necessarily unique. As it approaches **0**, this curve alternates between following the x-axis and following the y-axis for successive epochs.

In fact, one of the major open questions in geometric measure theory is whether an area-minimizing rectifiable current T has a unique oriented tangent cone at every point $a \in \mathrm{spt}\ T - \mathrm{spt}\ \partial T$.

Figure 9.7.3 illustrates the need for specifying that C be a cone.

9.8   Theorem [Federer, 5.4.3(6)]   *Let T be an area-minimizing rectifiable current in $\mathscr{R}_m$. Suppose $0 \in \operatorname{spt} T - \operatorname{spt} \partial T$. Then T has an oriented tangent cone C at 0.*

Proof   Consider a sequence $T_j = \mu_{r_j^{-1}\sharp} T$ of homothetic expansions of T with $r_j \to 0$. For $r_j \le r_0$ and fixed large $R > 0$, monotonicity of the mass ratio (9.3) states that

$$\mathbf{M}(T_j \llcorner \mathbf{B}(0, R)) = \mathbf{M}(T \llcorner B(0, Rr_j))r_j^{-m} \le \mathbf{M}(T \llcorner B(0, Rr_0))r_0^{-m} = c.$$

Also, for j large, T_j has no boundary inside $\mathbf{B}(0, R)$. Hence, by compactness (see Section 9.1), we may assume that T_j converges to a locally integral current C. The remark after 9.3 may be used to conclude that C is in fact a cone.   ∎

9.9   Theorem   *Let T be an area-minimizing rectifiable current in $\mathscr{R}_m$. Suppose $0 \in \operatorname{spt} T - \operatorname{spt} \partial T$. Let C be an oriented tangent cone to T at 0. Then $\Theta^m(C, 0) = \Theta^m(T, 0)$.*

Remark   Exercise 9.2 implies that C is itself area-minimizing.

Proof   After replacing the sequence r_j such that $\mu_{r_j^{-1}\sharp} T \to C$ with a subsequence if necessary, for each j choose currents A_j, B_j such that

$$\operatorname{spt}(\mu_{r_j^{-1}\sharp})T - (A_j + \partial B_j) \cap \mathbf{U}(0, 2) = \emptyset,$$

$$\mathbf{M}(A_j) + \mathbf{M}(B_j) \le 1/j^2.$$

Let $u(x) = |x - a|$ and apply slicing theory 4.11(4) to choose $1 < s_j < 1 + 1/j$ such that

$$\mathbf{M}\langle B_j, u, s_j + \rangle \le j\mathbf{M}(B_j) \le 1/j.$$

Note that

$$\mu_{r_j^{-1}\sharp} T \llcorner \mathbf{B}(0, s_j) = C \llcorner \mathbf{B}(0, s_j) + A_j \llcorner \mathbf{B}(0, s_j)$$

$$+ \ \partial(B_j \llcorner \mathbf{B}(0, s_j)) - \langle \mathbf{B}_j, u, s_j + \rangle.$$

Hence, $\mu_{r_j^{-1}\sharp} T \llcorner \mathbf{B}(0, s_j) \to C \llcorner \mathbf{B}(0, 1)$ in the flat norm. By the lower semicontinuity of mass (Exercise 4.22),

$$\Theta(C, 0) \le \Theta(T, 0).$$

Moreover, since $\mu_{r_j^{-1}\sharp} T \mathbf{B}(0, s_j)$ is area minimizing and

$$C \, \llcorner \, \mathbf{B}(0, s_j) + A_j \, \llcorner \, \mathbf{B}(0, s_j) - \langle B_j, u, s_j + \rangle$$

has the same boundary,

$$\mathbf{M}(\mu_{r_j^{-1}\sharp} T \, \llcorner \, \mathbf{B}(0, s_j)) \leq \mathbf{M}(C \, \llcorner \, \mathbf{B}(0, s_j)) + 2/j.$$

It follows that

$$\Theta(T, \, 0) \leq \Theta(C, \, 0). \qquad \blacksquare$$

Exercises

9.1 Give an example of an integral flat chain $T \in \mathscr{F}_0 \mathbf{R}^2$ such that $T \, \llcorner \, \mathbf{B}^2(0, 1)$ is not an integral flat chain.

9.2 Let $S_1, S_2, S_3, \ldots \to S$ be a convergent sequence of locally rectifiable currents. Suppose each S_j is area minimizing. Prove that S is area minimizing.

9.3 Let S be an area-minimizing rectifiable current in $\mathscr{R}_2 \mathbf{R}^3$ bounded by the circles $x^2 - y^2 = R^2$, $z = \pm 1$ oppositely oriented. Prove that

$$\operatorname{spt} S \subset \left\{ \sqrt{x^2 + y^2} \geq R - 2\sqrt{R} \right\}.$$

9.4 Prove or give a counterexample. If $T \in \mathbf{I}_2 \mathbf{R}^3$, then for all $a \in \operatorname{spt} T - \operatorname{spt} \partial T$, $\Theta^2(T, a) \geq 1$.

9.5 Let T be an m-dimensional area-minimizing rectifiable current in $\mathbf{R}^n$ and consider

$$f \colon \mathbf{R}^n \to \mathbf{R}, \quad f(x) = \Theta^m(T, x).$$

  a. Mention an example for which f is not continuous, even on $\operatorname{spt} T -$ $\operatorname{spt} \partial T$.

  b. Prove that f is upper semicontinuous on $\mathbf{R}^n - \operatorname{spt} \partial T$.

9.6 Let $T \in \mathscr{F}_m^{\mathrm{loc}}$, and let C be an oriented tangent cone to T at 0. Prove that spt $C \subset \operatorname{Tan}(\operatorname{spt} T, 0)$ (cf. 3.9). Show by example that equality need not hold.

9.7 Let $T \in \mathscr{F}_m^{\mathrm{loc}}$, and consider oriented tangent cones to T at 0: $C = \lim \mu_{r_j^{-1}\sharp} T$ and $D = \lim \mu_{s_j^{-1}\sharp} T$. Prove that, if $0 < \underline{\lim} s_j/r_j \leq \overline{\lim} s_j/r_j < \infty$, then $C = D$.

9.8 Let T be the polygonal curve which follows the x-axis from $1 = 2^{-0^2}$ to 2^{-1^2}, then the y-axis from 2^{-1^2} to 2^{-2^2}, and so on as in Figure 9.7.2, oriented outward.

    a. Show that the nonnegative x-axis and the nonnegative y-axis are each an oriented tangent cone at **0**.

    b. Find a limit of homothetic expansions which is not a cone.

    c. Let $T_0 = T \cap x$-axis. Show that the lower density $\Theta_*(T_0, \mathbf{0}) = 0$ while the analogous upper density $\Theta^*(T_0, \mathbf{0}) = 1/2$.

The Regularity of Area-Minimizing Hypersurfaces

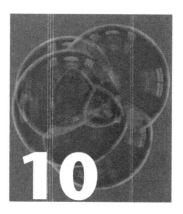

This chapter outlines some parts of the proof of the regularity theorem for area-minimizing rectifiable currents in $\mathscr{R}_{n-1}\mathbf{R}^n$ for $n \leq 7$. The purpose is to give an overview, illustrate basic arguments, and indicate why regularity fails for $n \geq 8$. The deeper and more technical aspects of the theory are omitted. The first theorem proves a special case by methods that will be useful in the general case.

10.1 Theorem *Let T be an area-minimizing rectifiable current in $\mathscr{R}_1\mathbf{R}^2$. Then* spt $T -$ spt ∂T *consists of disjoint line segments.*

Proof It will be shown that every point $a \in$ spt $T -$ spt ∂T has a neighborhood $\mathbf{U}(a, r)$ such that spt $T \cap \mathbf{U}(a, r)$ is a straight line segment.

CASE 1 *If ∂T consists of two points (oppositely oriented), then T is the oriented line segment between them.* Our assignment is to prove the most famous result in the calculus of variations: that a straight line is the shortest distance between two points! We may assume $\partial T = \delta_{(1,0)} - \delta_{(0,0)}$. Let T_0 be the oriented segment from $(0, 0)$ to $(1, 0)$:

$$T_0 = [(0, 0), (1, 0)] = \mathscr{H}^1 \llcorner \{0 \leq x \leq 1, \, y = 0\} \wedge \mathbf{i}.$$

We will actually show that T_0 uniquely minimizes length among all normal currents $N \in \mathbf{N}_1\mathbf{R}^2$ with the same boundary as T_0. Indeed,

$$\mathbf{M}(N) \geq N(dx) = \partial N(x) = 1 = \mathbf{M}(T_0).$$

Therefore, T_0 is area minimizing. Furthermore, if $\mathbf{M}(N) = 1$, then $\vec{N} = \mathbf{i} \, ||N||$-almost everywhere.

Next, supposing that $\mathbf{M}(N) = 1$, we show that spt $N \subset \{y = 0\}$. If not, for some $\varepsilon > 0$ there is a C^∞ function $0 \leq f(y) \leq 1$ such that $f(y) = 1$ for $|y| \leq \varepsilon$ and $\mathbf{M}(N \llcorner f) < 1$.

Geometric Measure Theory. http://dx.doi.org/10.1016/B978-0-12-804489-6.00010-1

$$\partial(N \llcorner f) = (\partial N) \llcorner f - N \llcorner df = \partial N - 0,$$

because $\vec{N} = \mathbf{i} \, ||N||$-almost everywhere and $df(\mathbf{i}) = 0$. Since $\partial(N \llcorner f) = \partial N = \partial T_0$ and T_0 is mass minimizing, therefore $\mathbf{M}(N \llcorner f) \geq 1$. This contradiction proves that spt $N \subset \{y = 0\}$.

Finally, note that $\partial(N - T_0) = 0$. By the constancy theorem, 4.9, $N - T_0$ is a multiple of $\mathbf{E}^1 \equiv \mathcal{H}^1 \wedge \mathbf{i}$. Since $N - T_0$ has compact support, it must be 0. Therefore, $N = T_0$, uniqueness is proved, and Case 1 is complete.

CASE 2   *If the density $\Theta^1(T, a)$ equals 1, then* spt T *is a straight line segment in some neighborhood* $\mathbf{U}(a, r)$ *of a.* For almost all s, $0 < s < \text{dist}(a, \text{spt } \partial T)$, the slice $\partial(T \llcorner \mathbf{B}(a, s))$ is a zero-dimensional rectifiable current and a boundary—that is, an even number of points (counting multiplicities). There cannot be 0 points because then $T' = 0$ would have the same boundary and less mass than $T \llcorner \mathbf{B}(a, s)$. Therefore, $\mathbf{M}(\partial(T \llcorner \mathbf{B}(a, s))) \geq 2$. On the other hand, by slicing theory 4.11(4),

$$s^{-1} \int_0^s \mathbf{M}(\partial(T \llcorner \mathbf{B}(a, r))) \, dr \leq s^{-1} \mathbf{M}(T \llcorner \mathbf{B}(a, s)),$$

which converges to $\alpha_1 \Theta^1(T, a) = 2$ as $s \to 0$. Therefore, for some small $r > 0$, $\mathbf{M}(\partial(T \llcorner \mathbf{B}(a, r))) = 2$, and $\partial(T \llcorner \mathbf{B}(a, r))$ consists of two points. By Case 1, $\text{spt}(T \llcorner \mathbf{B}(a, r))$ is a line segment, as desired.

The general case will require the following lemma.

Lemma [Federer, 4.5.17]   *If $R \in \mathcal{R}_{n-1} \mathbf{R}^n$ with $\partial R = 0$, then there are nested, $\mathcal{L}^n$ measurable sets M_i ($i \in \mathbf{Z}$), $M_i \subset M_{i-1}$, such that*

$$R = \sum_{i \in \mathbf{Z}} \partial(\mathbf{E}^n \llcorner M_i) \text{ and } \mathbf{M}(R) = \sum_{i \in \mathbf{Z}} \mathbf{M}(\partial(\mathbf{E}^n \llcorner M_i)).$$

Here, $\mathbf{E}^n$ is the unit, constant-coefficient n-dimensional current in $\mathbf{R}^n$, defined by

$$\mathbf{E}^n = \mathcal{L}^n \wedge \mathbf{e}_1 \wedge \cdots \wedge \mathbf{e}_n.$$

Proof of Lemma   By the isoperimetric inequality (5.3), $R = \partial T$ for some $T \in \mathbf{I}_n \mathbf{R}^n$. Such a T is of the form $T = \mathbf{E}^n \llcorner f$ for some measurable, integer-valued function f, and $\mathbf{M}(T) = \int |f|$. Just put $M_i = \{x : f(x) \geq i\}$. See Figure 10.1.1. All of the conclusions of the lemma except the last follow immediately. The last conclusion on $\mathbf{M}(R)$ means that there is no cancellation in the sum $R = \sum \partial(\mathbf{E}^n \llcorner M_i)$. The idea of the proof is that because the M_i are nested, their boundaries, if they happen to overlap, have similar orientations. Hence, in their sum, the masses add. We omit the details.

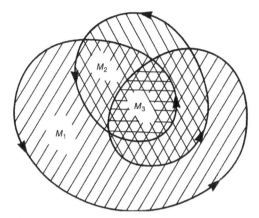

Figure 10.1.1. Decomposing this curve as the boundaries of three nested regions yields three pieces which never cross over each other.

CASE 3   GENERAL CASE   *For every* $a \in$ spt $T-$ spt ∂T, spt T *is a straight line segment in some neighborhood* $\mathbf{U}(a, r)$ *of* a. Choose $0 < \rho < \text{dist}(a, \text{spt } \partial T)$ such that $\mathbf{M}(\partial(T\llcorner\mathbf{B}(a, \rho))) < \infty$. Let Ξ be a rectifiable current supported in the sphere $\mathbf{S}(a, p)$ with $\partial \Xi = \partial(T\llcorner\mathbf{B}(a, \rho))$ (Figure 10.1.2). Apply the previous lemma to $T\llcorner\mathbf{B}(a, \rho) - \Xi$, and let $T_i = (\partial(\mathbf{E}^2\llcorner M_i))\llcorner\mathbf{U}(a, \rho)$. Since $T\llcorner\mathbf{U}(a, \rho) = \sum T_i$ and $\mathbf{M}(T\llcorner\mathbf{U}(a, \rho)) = \sum \mathbf{M}(T_i)$, each T_i is mass minimizing. Since $\mathbf{M}(T\llcorner\mathbf{U}(a, \rho)) < \infty$, it follows from monotonicity (9.5 and 9.6) that spt T_i intersects $\mathbf{U}(a, \rho/2)$ for only finitely many i, and we will ignore the rest. Since T_i is of the form $(\partial(\mathbf{E}^2\llcorner M_i))\llcorner\mathbf{U}(a, \rho)$, it can be shown with some work that an oriented tangent cone C to T_i at any point in spt $T_i \cap \mathbf{U}(a, \rho)$ is of the form $C = \partial(\mathbf{E}^2\llcorner N)$ (cf. Federer [1, 5.4.3]). Similarly, if $b \in$ spt $C - \{0\}$, then an oriented tangent cone D to C at b is of the form $D = \partial(\mathbf{E}^2\llcorner P)$. The fact that C is a cone means that D is a *cylinder* in the sense of invariance under translations in the b direction. A relatively easy argument shows that a one-dimensional oriented cylinder is an oriented line with multiplicity (cf. Federer [1, 4.3.15]). Since D is of the form $\partial(\mathbf{E}^2\llcorner P)$, the multiplicity must be 1. Consequently, $\Theta^1(C, b) = \Theta^1(D, \mathbf{0}) = 1$, for all $b \in$ spt $C-\{0\}$. By Case 2, C consists of rays emanating from **0**. Since $C = \partial(\mathbf{E}^2\llcorner N)$ has no boundary, the same number must be oriented outward as inward. Since C is area minimizing, oppositely oriented rays must be at a 180° angle; otherwise, inside the unit circle they could be replaced by a straight line of less mass. See Figure 10.1.3. Therefore, C must be an oriented line with multiplicity. Since C is of the form $C = \partial(\mathbf{E}^2\llcorner N)$, the multiplicity must be 1. Consequently, $\Theta^1(T_i, a) = \Theta^1(C, \mathbf{0}) = 1$. By Case 2, for some $0 < r_i \leq \rho/2$, $T_i\llcorner\mathbf{U}(\mathbf{0}, r_i) = (\partial(\mathbf{E}^2\llcorner M_i))\llcorner\mathbf{U}(\mathbf{0}, r_i)$ is an oriented line. Let r be the least of the finitely many r_i. Since $M_i \subset M_{i-1}$, the various nonzero $T_i\llcorner\mathbf{U}(\mathbf{0}, r)$ coincide. Therefore, $T\llcorner\mathbf{U}(\mathbf{0}, r)$ is an oriented line with multiplicity.   ∎

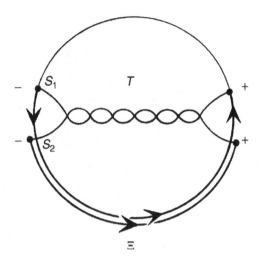

Figure 10.1.2. Given the surface $T \llcorner \mathbf{B}(a, \rho)$ in the ball with boundary in the sphere, there is a surface Ξ entirely in the sphere with the same boundary.

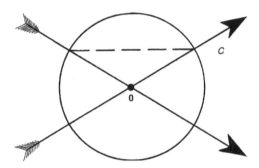

Figure 10.1.3. A pair of lines through **0** cannot be mass minimizing, as the dashed short-cut shows.

Now we state and sketch the proof of a complete interior regularity theorem for area-minimizing rectifiable currents in $\mathbf{R}^n$, for $n \leq 7$.

10.2 Regularity for Area-Minimizing Hypersurfaces Theorem (Simons; see Federer [1, 5.4.15]) *Let T be an area-minimizing rectifiable current in $\mathscr{R}_{n-1} \mathbf{R}^n$ for $2 \leq n \leq 7$. Then* spt $T -$ spt ∂T *is a smooth, embedded manifold.*

The proof depends on a deep lemma, which we will not prove. Our intent is to show how the pieces fit together. This lemma gives conditions on density and tangent cones sufficient to establish regularity. A later regularity theorem of Allard [Section 8] (see Simon [Chap. 5]) shows that the hypothesis on an oriented tangent cone is superfluous. Moreover, Allard weakens the "area-minimizing" hypothesis to "weakly bounded mean curvature."

10.3   Lemma [Federer, 5.4.6]   *For* $1 \leq m \leq n - 1$, *suppose* T *is an area-minimizing locally rectifiable current in* $\mathscr{R}_m^{\text{loc}}$ $\mathbf{R}^n$, $a \in \text{spt } T -$ *spt* ∂T, $\Theta^m(T, a) = 1$, *and some oriented tangent cone to* T *at* a *is an oriented, m-dimensional plane. Then* spt T *is a smooth, embedded manifold at* a.

The second lemma gives a maximum principle for area-minimizing hypersurfaces. The result is contained in Federer [5.3.18], but it is easier to prove.

10.4   Maximum Principle   *For* $n \geq 2$, *let* S_1 *and* S_2 *be* $(n - 1)$-*dimensional, area-minimizing rectifiable currents in* $\mathbf{R}^n$, *and let* $M_i = \text{spt } S_i$. *Suppose that* M_1 *and* M_2 *intersect at a point* a; *that, in some neighborhood of* a, M_1 *and* M_2 *are smooth submanifolds; and that* M_2 *lies on one side of* M_1. *Then in some neighborhood of* a, M_1 *and* M_2 *coincide.*

Remark   The maximum principle holds for singular area-minimizing hypersurfaces and more generally. Indeed, Wickramasekera generalizes it to minimal hypersurfaces (stationary integral varifolds, not necessarily area minimizing) one of whose singular sets has $(n - 2)$-dimensional Hausdorff measure 0. Solomon and White treat hypersurfaces stationary for smooth, even elliptic integrands, at least one of which is smooth.

Proof   At a, M_1 and M_2 may be viewed locally as graphs of functions u_1 and u_2, satisfying the minimal surface equation (6.2(1)). By a standard maximum principle, due essentially to Hopf [Satz 1'], the functions u_1 and u_2 coincide. ∎

The next lemma of Simons provided the final ingredient for the regularity theorem. In a remark following the proof of the theorem we discuss its failure for $n \geq 8$. This lemma marks a solo appearance of differential geometry without measure theory.

10.5   Simons's Lemma [Federer, 5.4.14]   *For* $3 \leq n \leq 7$, *let* B *be an oriented, compact,* $(n - 2)$-*dimensional smooth submanifold of the unit* $(n - 1)$-*dimensional sphere, such that the cone over* B *is area minimizing. Then* B *is a great sphere.*

It is customarily assumed that the submanifold B is connected. That weaker version of the lemma, applied to the connected components of a general submanifold B, shows that all components are great spheres and hence intersect. This contradiction shows that B is connected and renders that assumption superfluous.

The final lemma has an easy proof by slicing.

10.6 Lemma [Federer, 5.4.8, 5.4.9] *For $1 \leq m \leq n - 1$, let Q be a locally rectifiable current in $\mathscr{R}_m^{\mathrm{loc}} \mathbf{R}^n$. Then Q is area minimizing if and only if $\mathbf{E}^1 \times Q$ is area minimizing.*

Proof of Theorem 10.2 The proof will be in two parts, by induction. The initial case, $n = 2$, was proved by Theorem 10.1.

PART 1 *Suppose S is an area-minimizing rectifiable current in $\mathscr{R}_{n-1} \mathbf{R}^n$ and S is of the form $S = (\partial (\mathbf{E}^n \llcorner M)) \llcorner V$ for some measurable set M and open set V. Then* spt $S \cap V$ *is a smooth embedded manifold.* To prove Part 1, let $a \in$ spt $S \cap V$. An oriented tangent cone C to S at a can be shown to be of the form $C = \partial (\mathbf{E}^n \llcorner N)$ (cf. Federer [1, 5.4.3]). Similarly, for $b \in$ spt $C - \{0\}$, an oriented tangent cone D to C at b is of the form $D = \partial (\mathbf{E}^n \llcorner P)$. The fact that C is an oriented cone means that D is an "oriented cylinder" of the form $D = \mathbf{E}^1 \times Q$ for some $Q \in \mathscr{R}_{n-2} \mathbf{R}^{n-1}$ (cf. Federer [1, 4.3.15]). Since S is area minimizing, so are C and $D = \mathbf{E}^1 \times Q$. By Lemma 10.6 and induction, Q is an oriented, smooth, embedded manifold, possibly with multiplicity. Hence, D is an oriented, smooth, embedded manifold, with multiplicity 1 because D is of the form $D = \partial (\mathbf{E}^n \llcorner P)$. Therefore, any oriented tangent cone to D at $\mathbf{0}$ is an oriented, $(n-1)$-dimensional plane. In particular, since D is an oriented cone, D is an oriented, $(n - 1)$-dimensional plane. Therefore, by Lemma 10.3, spt $C - \{0\}$ is a smooth, embedded manifold. By Lemma 10.5, spt C intersects the unit sphere in a great sphere. Hence, C is an oriented, $(n - 1)$-dimensional plane with multiplicity 1 because C is of the form $C = \partial (\mathbf{E}^n \llcorner N)$. A reapplication of Lemma 10.3 now shows that spt S is a smooth, embedded manifold at a, proving Part 1.

PART 2 *Completion of proof.* Let $a \in$ spt $T -$ spt ∂T. Choose a small $\rho > 0$ such that $\partial (T \llcorner \mathbf{U}(a, \rho))$ is rectifiable (cf. 4.11(5)). Let Ξ be a rectifiable current supported in the sphere $\mathbf{S}(a, p)$ with the same boundary as $T \llcorner \mathbf{U}(a, \rho)$. By the Lemma of section 10.1, there are nested sets $M_i \subset M_{i-1}$ such that $T \llcorner \mathbf{U}(a, \rho) = \sum S_i$, with $S_i = (\partial (\mathbf{E}^n \llcorner M_i)) \llcorner \mathbf{U}(a, \rho)$. Moreover, $\mathbf{M}(T \llcorner \mathbf{U}(a, \rho)) = \sum \mathbf{M}(S_i)$ so that each S_i is area minimizing. Since $\mathbf{M}(T \llcorner \mathbf{U}(a, \rho)) < \infty$, it follows from monotonicity (9.3 and 9.6) that spt S_i intersects $\mathbf{U}(a, \rho/2)$ for only finitely many i. For such i, by Part I, spt S_i is a smooth embedded manifold at a. The containments $M_i \subset M_{i-1}$ imply that each of these manifolds lies on one side of another. Therefore, by the maximum principle, 10.4, spt T is a smooth, embedded manifold at a. The theorem is proved.

10.7 Remarks The regularity theorem fails for $n \geq 8$. E. Bombieri, E. De Giorgi, and E. Giusti (1969) gave an example of a seven-dimensional, area-minimizing rectifiable current T in $\mathbf{R}^8$ with an isolated singularity at $\mathbf{0}$. This current T is the oriented truncated cone over $B = \mathbf{S}^3(\mathbf{0}, 1/\sqrt{2}) \times \mathbf{S}^3 (\mathbf{0}, 1/\sqrt{2}) \subset \mathbf{S}^7(\mathbf{0}, 1) \subset \mathbf{R}^8; \partial T = B$. It also provides a counterexample to Simons's Lemma 10.5, which is precisely the point at which the proof of regularity breaks down.

Here, we give a plausibility argument that such a counterexample should arise for some large n. First, consider $B = \mathbf{S}^0(\mathbf{0}, 1/\sqrt{2}) \times \mathbf{S}^0(\mathbf{0}, 1/\sqrt{2}) \subset \mathbf{S}^1 \subset \mathbf{R}^2$, the four points shown in Figure 10.7.1. The associated cone does have a singularity at $\mathbf{0}$, but it is not mass minimizing. The mass-minimizing current consists of two vertical lines (or two horizontal lines).

Second, consider $B = \mathbf{S}^1(\mathbf{0}, 2/\sqrt{5}) \times \mathbf{S}^0(\mathbf{0}, 1/\sqrt{5}) \subset \mathbf{S}^2 \subset \mathbf{R}^3$, the two circles shown in Figure 10.7.2. Again, the associated cone has a singularity at $\mathbf{0}$, but it is not mass minimizing. The mass-minimizing current is the pictured catenoid. The catenoid has less area than a cylinder. Although the curved cross sections are longer than the straight lines of the cylinder, the circumference is less. The amount that the catenoid bows inward toward the cone is a balancing of these two effects.

Figure 10.7.1. The X-shape, which is the cone over the four points $\mathbf{S}^0 \times \mathbf{S}^0$ in $\mathbf{R}^2$, is not mass minimizing. The mass-minimizing current consists of two vertical line segments.

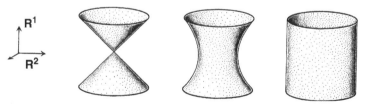

Figure 10.7.2. The cone bounded by the two circles $\mathbf{S}^1 \times \mathbf{S}^0$ in $\mathbf{R}^3$ is not mass minimizing. Neither is the cylinder on the right. The mass-minimizing surface is the catenoid in the middle. It bows inward slightly toward the cone in order to shorten its waist but not too much to overstretch its sides.

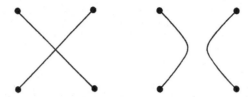

Figure 10.7.3. In higher dimensions, the mass-minimizing surface bows further inward toward the cone. Finally, in $\mathbf{R}^8$ it collapses onto the cone, which is mass minimizing for the first time.

Third, consider $B = \mathbf{S}^2(0, \ 1/\sqrt{2}) \times \mathbf{S}^2(0, \ 1/\sqrt{2}) \subset \mathbf{S}^5 \subset \mathbf{R}^6$, shown schematically in Figure 10.7.3. As the dimensions increase, the mass cost of being far from the origin rises, and the mass-minimizing current bows farther toward the cone. Finally, in $\mathbf{R}^8$, it has collapsed onto the cone, which is mass minimizing for the first time.

Exercises

10.1 Give two different area-minimizing rectifiable currents in $\mathscr{R}_2 \mathbf{R}^3$ with the same boundary.

10.2 Prove or give a counterexample: If $T \in \mathscr{R}_m\{\mathbf{R}^n \times \mathbf{R}^l\}$ is area minimizing and p denotes orthogonal projection of $\mathbf{R}^n \times \mathbf{R}^l$ onto $\mathbf{R}^n$, then $p_\sharp T$ is area minimizing.

10.3 Find a counterexample in $\mathbf{I}_2 \mathbf{R}^3$ to Lemma 10.3 if the hypothesis that S be area minimizing is removed.

Flat Chains Modulo v, Varifolds, and (M, ε, δ)-Minimal Sets

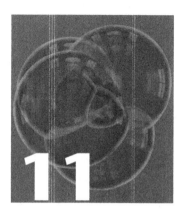

A number of alternative spaces of surfaces have been developed in geometric measure theory, as required for theory and applications. This chapter gives brief descriptions of flat chains modulo v, varifolds, and (**M**, ε, δ)-minimal sets.

11.1 Flat Chains Modulo v [Federer, 4.2.26] One way to treat nonorientable surfaces and more general surfaces is to work modulo 2, or more generally modulo v for any integer $v \geq 2$. Two rectifiable currents T_1, T_2 are *congruent modulo v* if $T_1 - T_2 = vQ$ for some rectifiable current Q. In particular, $T \equiv -T$ (mod 2). The m-dimensional rectifiable currents modulo v, denoted $\mathscr{R}_m^v$, are defined as congruence classes of rectifiable currents.

For example, consider the Möbius strip of Figure 11.1.2 bounded by the curve C. There is no way to orient it to turn it into a rectifiable current with boundary C. However, if it is cut along a suitable curve D, the Möbius strip can then be oriented as a rectifiable current T, and it works out that $\partial T \equiv C$ (mod 2). In general, rectifiable currents modulo 2 correspond to unoriented surfaces.

Two parallel circles, as in Figure 11.1.3, bound an interesting rectifiable current modulo 3. The surfaces of both Figures 11.1.2 and 11.1.3 occur as soap films.

Most of the concepts and theorems on rectifiable currents have analogs for rectifiable currents modulo v: mass, flat norm, the deformation theorem, the compactness theorem, existence theory, and the approximation theorem.

One tricky point: to ensure completeness, the integral flat chains congruent to 0 modulo v must be defined as the *flat-norm closure* of those of the form vQ. In codimension greater than 1, it is an open question whether they are all of the form vQ. (The counterexample in Federer [p. 426] is wrong.) However, every rectifiable current congruent to 0 modulo v is of the form vQ.

Geometric Measure Theory. http://dx.doi.org/10.1016/B978-0-12-804489-6.00011-3

Figure 11.1.1. Bill Ziemer (right), who introduced flat chains modulo 2, with his thesis advisor, Wendell Fleming (left), and the author (center), at a celebration in Ziemer's honor at Indiana in 1994. Photo courtesy of Ziemer.

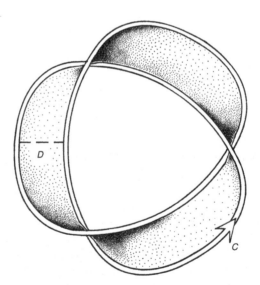

Figure 11.1.2. A rectifiable current modulo 2 can be nonorientable, like this Möbius strip.

It is generally easier to prove regularity for area-minimizing rectifiable currents modulo 2 than for rectifiable currents themselves. Indeed, an m-dimensional area-minimizing rectifiable current modulo 2 in $\mathbf{R}^n$ is a smooth embedded manifold on the interior, except for a rectifiable singular set of locally finite

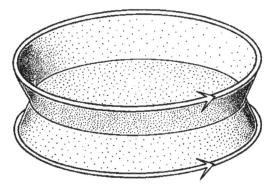

Figure 11.1.3. In a rectifiable current modulo 3, three sheets can meet along a curve inside the surface which does not count as boundary.

$\mathcal{H}^{m-2}$ measure (Federer [2]; Simon [1]). The only standard regularity result I know which fails modulo 2 is boundary regularity for area-minimizing hypersurfaces (8.4).

For $\nu > 2$, even area-minimizing hypersurfaces modulo ν can have codimension 1 singular sets, as Figure 11.1.3 suggests. Taylor [3] proved that a two-dimensional area-minimizing rectifiable current modulo 3 in $\mathbf{R}^3$ away from boundary consists of C^∞ surfaces meeting in threes at 120° angles along $C^{1,\alpha}$ curves (cf. 13.9). White [8] proved that an $(n-1)$-dimensional area-minimizing rectifiable current modulo 4 in $\mathbf{R}^n$ decomposes locally into a pair of area-minimizing rectifiable currents modulo 2. Moreover, White [7] proved that for any odd ν, an $(n-1)$-dimensional area-minimizing rectifiable current modulo ν in $\mathbf{R}^n$ away from boundary is a smooth embedded manifold, except for a singular set of dimension at most $n-2$. For even $\nu > 2$, almost everywhere regularity remains an open question.

11.2 Varifolds [Allard]

Varifolds provide an alternative perspective to currents for working with rectifiable sets. Varifolds carry no orientation, and hence there is no cancellation in the limit and no obvious definition of boundary. An m-dimensional *varifold* is a Radon measure on $\mathbf{R}^n \times G_m\mathbf{R}^n$, where $G_m\mathbf{R}^n$ is the Grassmannian of unoriented unit m-planes through $\mathbf{0}$ in $\mathbf{R}^n$. We have seen previously how to associate to a rectifiable set E the measure $\mathcal{H}^m \llcorner E$ on $\mathbf{R}^n$, but this perspective ignores the tangent planes to E. Instead, we now associate to a rectifiable set E, with unoriented tangent planes $\vec{E}(x) = \text{Tan}^m(E, x) \in G_m\mathbf{R}^n$, the varifold

$$\mathbf{v}(E) \equiv \mathcal{H}^m \llcorner \{(x, \vec{E}(x)) : x \in E\}.$$

The varifolds which so arise are called *integral varifolds*. One allows positive integer multiplicities and noncompact support.

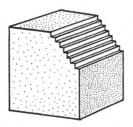

Figure 11.2.1. If you cut an edge off a cubical crystal, the exposed surface forms tiny steps, well modeled by a varifold.

In general varifolds, the tangent planes need not be associated with the underlying set. For example, if you cut an edge off a cubical crystal, as in Figure 11.2.1, because the crystal loves horizontal and vertical directions, the exposed surface forms very small horizontal and vertical steps. Such a corrugation can be modeled by a varifold concentrated half and half on horizontal and vertical tangent planes in $G_2\mathbf{R}^3$, whereas in space $\mathbf{R}^3$ it is concentrated on the diagonal surface.

The *first variation* δV of a varifold V is a function which assigns to any compactly supported smooth vector field g on $\mathbf{R}^n$ the initial rate of change of the area of V under a smooth deformation of $\mathbf{R}^n$ with initial velocity g. Roughly, the first variation is due to the mean curvature of V and the boundary of V. A varifold is called *stationary* if $\delta V = 0$. Geometrically, stationary integral varifolds include area-minimizing rectifiable currents, area-minimizing rectifiable currents modulo ν, and many other physical surfaces such as soap films. Some singularities of soap films unavoidably count as additional boundary in the category of rectifiable currents or rectifiable currents modulo ν, but fortunately, they do not add to their first variations as varifolds.

Integral varifolds of bounded first variation include surfaces of constant or bounded mean curvature and soap bubble clusters. They satisfy a weakened version of Monotonicity 9.3: *the area ratio times e^{Cr} is monotonically increasing*, where C is a bound on the first variation or mean curvature.

There is a compactness theorem for integral varifolds with bounds on their areas, first variations, and supports. There are also general isoperimetric and regularity theorems. However, it is an open question whether a two-dimensional stationary integral varifold in an open subset of $\mathbf{R}^3$ is a smooth embedded manifold almost everywhere.

11.3 (M, ε, δ)-Minimal Sets [Almgren 1]

Perhaps the best model of soap films is provided by the $(\mathbf{M}, 0, \delta)$-minimal sets of Almgren. A nonempty, bounded subset $S \subset \mathbf{R}^n - B$ with $\mathscr{H}^m(S) < \infty$ and $S = \mathrm{spt}(\mathscr{H}^m \llcorner S) - B$ is $(\mathbf{M}, 0, \delta)$ *minimal* with respect to a closed set B (typically "the boundary") if, for every Lipschitz deformation φ of $\mathbf{R}^n$ which differs from the identity map only in a δ-ball disjoint from B,

$$\mathscr{H}^m(S) \leq \mathscr{H}^m(\varphi(S)).$$

Since φ need not be a diffeomorphism, it can pinch pieces of surface together, as in Figure 11.3.1. The **M** refers to area and may be replaced by a more general integrand. For more general functions $\varepsilon(r) = Cr^\alpha$, $\alpha > 0$, the inequality imposed on deformations inside r-balls ($r \leq \delta$) relaxes to

$$\mathscr{H}^m(S) \leq (1 + \varepsilon(r))\mathscr{H}^m(\varphi(S)).$$

Such (**M**, ε, δ)-minimal sets include soap bubbles (with volume constraints); see Section 13.8.

Three basic properties of (**M**, ε, δ)-minimal sets are ($\mathscr{H}^m$, m) rectifiability [Almgren 1, II.3(9)], monotonicity [Taylor 4, II.1], and the existence of an (**M**, 0, δ)-minimal tangent cone at every point [Taylor 4, II.2]. The condition in the definition that $S = \mathrm{spt}(\mathscr{H}^m \llcorner S) - B$ may be replaced by the condition that S be rectifiable, with the understanding that S may be altered by a set of $\mathscr{H}^m$ measure 0 [Morgan 17, §2.5].

Almgren [1] has proved almost-everywhere regularity results for (**M**, ε, δ)-minimal sets. In 1976, Taylor [4] proved that for two-dimensional (**M**, ε, δ)-minimal sets in $\mathbf{R}^3$, there are only two possible kinds of singularities: (1) three sheets of surface meeting at 120° angles along a curve and (2) four such curves meeting at approximately 109° angles at a point (see Section 13.9). These are precisely the two kinds of singularities that Plateau had observed in soap bubbles and soap films 100 years earlier. (Warning: Almgren [1] and Taylor [4] are technical, and I think there are some (correctable) gaps.)

Boundary regularity remains conjectural; see Section 13.9.

It remains an open question whether a smooth curve in $\mathbf{R}^3$ bounds a least-area soap film in the class of (**M**, 0, δ)-minimal sets, with variable $\delta > 0$. Minimizing area for fixed δ follows e.g. from compactness for stationary integral varifolds, of which (**M**, 0, δ)-minimal sets are trivially a closed subset. The problem is that

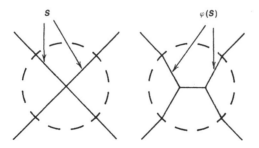

Figure 11.3.1. A curve S is not (**M**, 0, δ) minimal if a deformation $\phi(S)$ has less length.

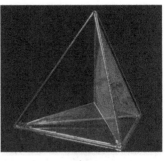

Figure 11.3.2. It is an open question whether the cone over the tetrahedron is a least-area soap film or whether there might be a smaller soap film of higher topological type. Photo by F. Goro and computer graphics by J. Taylor; used by permission.

in the limit δ may go to 0. See Morgan [20]. It is an open question whether the cone over the tetrahedron (the soap film of Figure 11.3.2) is a least-area soap film or whether there might be a smaller soap film of higher topological type. Another open question asks whether the Cartesian product of an $(\mathbf{M}, 0, \delta)$-minimal set with an interval is $(\mathbf{M}, 0, \delta')$ minimal (a property which holds trivially for most classes of minimal surfaces; cf. 10.6).

Exercises

11.1 Give an example of a boundary curve in $\mathbf{R}^3$ for which the area-minimizing flat chain modulo 4 has less area than the area-minimizing integral current.

11.2 Let S be the unit 2 disc, and let $\mathbf{v}(S)$ be the associated varifold. What is $\mathbf{v}(S)(\mathbf{R}^n \times G_2\mathbf{R}^n)$?

11.3 Give an example of a two-dimensional set in $\mathbf{R}^3$ that is $(\mathbf{M}, 0, \delta)$ minimal for small δ but not for large δ.

11.4 Give an example of a two-dimensional set in $\mathbf{R}^3$ that is $(\mathbf{M}, 0, \delta)$ minimal for all $d > 0$ but not area minimizing.

Miscellaneous Useful Results

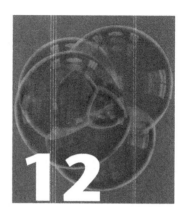

Federer's treatise presents many basic methods of geometry and analysis in a generality that embraces manifold applications. This chapter describes Federer's treatment of Sard's theorem, Green's theorem, relative homology, functions of bounded variation, and general parametric integrands.

12.1 Morse–Sard–Federer Theorem

The usual statement of Sard's theorem says that the set of critical values of a C^∞ function $f : \mathbf{R}^m \to \mathbf{R}^n$ has Lebesgue measure 0. Federer's refinement shows precisely how the Hausdorff measure of the image depends on the rank of Df and the smoothness class of f.

Theorem [Federer, 3.4.3] *For integers* $m > \nu \geq 0, k \geq 1$, *let* f *be a* C^k *function from an open subset* A *of* $\mathbf{R}^m$ *into a normed vectorspace* Y. *Then*

$$\mathscr{H}^{\nu + \frac{m-\nu}{k}} f(\{x \in A : \operatorname{rank} Df(x) \leq \nu\}) = 0.$$

Note that the usual statement may be recovered by taking $Y = \mathbf{R}^n$, $\nu = n - 1$, $k \geq m - n + 1$. The latest improvement has been provided by Bates.

12.2 Gauss–Green–De Giorgi–Federer Theorem

The usual statement of Green's theorem says that a C^1 vector field $\xi(x)$ on a compact region A in $\mathbf{R}^n$ with C^1 boundary B satisfies

$$\int_B \xi(x) \cdot \mathbf{n}(A, x) \, d\sigma = \int_A \operatorname{div} \xi(x) \, d\mathscr{L}^n x,$$

Geometric Measure Theory. http://dx.doi.org/10.1016/B978-0-12-804489-6.00012-5

111

where $\mathbf{n}(A, x)$ is the exterior unit normal to A at x and $d\sigma$ is the element of area on B. Federer treats more general regions and vector fields that are merely Lipschitz.

Federer allows measurable regions A for which the current boundary $T = \partial(\mathbf{E}^n \llcorner A)$ is representable by integration: $T(\varphi) = \int \langle \vec{T}, \varphi \rangle \, d \|T\|$ (cf. 4.3). If A is compact, this condition just says that the current boundary has finite measure: $\mathbf{M}(T) = \|T\| (\mathbf{R}^n) < \infty$. In any case, this condition is weaker than requiring that the topological boundary of A have finite $\mathcal{H}^{n-1}$ measure.

Definition Let $b \in A \subset \mathbf{R}^n$. We call $\mathbf{n} = \mathbf{n}(A, b)$ the *exterior normal of A at b* if $\mathbf{n}$ is a unit vector,

$$\Theta^n(\{x : (x - b) \cdot \mathbf{n} > 0\} \cap A, \, b) = 0,$$

and

$$\Theta^n(\{x : (x - b) \cdot \mathbf{n} < 0\} - A, \, b) = 0.$$

Clearly, there is at most one such $\mathbf{n}$. If b is a smooth boundary point of A, then $\mathbf{n}$ is the usual exterior normal. Even if ∂A is not smooth at b, $\mathbf{n}$ may be defined, as Figure 12.2.1 suggests. The assertion in the following theorem that the measure $\|T\| = \mathcal{H}^{n-1} \llcorner$ domain $\mathbf{n}(A, x)$ says roughly that the current boundary of A coincides with the domain of $\mathbf{n}(A, x)$ almost everywhere. In the final formula, div ξ exists almost everywhere because a Lipschitz function is differentiable almost everywhere.

Theorem [Federer, 4.5.6, 4.5.12] *Let A be an $\mathcal{L}^n$-measurable subset of $\mathbf{R}^n$ such that $T = \partial(\mathbf{E}^n \llcorner A)$ is representable by integration (which holds, e.g., if the $\mathcal{H}^{n-1}$ measure of the topological boundary of A is finite). Then $\|T\| = \mathcal{H}^{n-1} \llcorner$ domain $\mathbf{n}(A, x)$ and, for any Lipschitz vector field $\xi(x)$ of compact support,*

$$\int \xi(x) \cdot \mathbf{n}(A, \, x) \, d \, \mathcal{H}^{n-1} x = \int_A \mathrm{div} \, \xi(x) \, d\mathcal{L}^n \, x.$$

Generalizations to wilder sets have been provided by Harrison. See De Pauw.

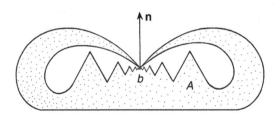

Figure 12.2.1. The generalized normal $\mathbf{n}$ is defined at b because the arms from the sides have density 0 at b.

12.3   Relative Homology [Federer, 4.4] Suppose $B \subset M$ are C^1,
compact submanifolds with boundary of $\mathbf{R}^n$ (or, more generally, compact Lipschitz neighborhood retracts; cf. Federer [1, 4.1.29, 4.4.1, 5.1.6]). Two rectifiable currents S, T in M are *homologous with respect to* $\mathbf{B}$ if there is a rectifiable current X in M such that

$$\operatorname{spt}(T - S - \partial X) \subset B.$$

We say that S and T are in the same relative homology class. Given a rectifiable current S, there is a rectifiable current T of least area in its relative homology class.

Example   Let M be a perturbed solid torus in $\mathbf{R}^3$, let B be its boundary, and let S be a cross-sectional disc. The area-minimizer T relatively homologous to S provides a cross-sectional surface of least area. See Figure 12.3.1. The boundary of T is called a *free boundary*.

Example   Let M be a large, encompassing ball in $\mathbf{R}^3$, let B be the surface of a table (not necessarily flat), and let C be a curve which begins and ends in B. Let S be a surface with $\operatorname{spt}(\partial S - C) \subset B$. The area-minimizer T relatively homologous to S provides a surface of least area with the fixed boundary C and additional free boundary in B. It can generally be realized as a soap film. See Figure 12.3.2.

Proof of Existence   As Brian White pointed out to me, there is a much simpler existence proof than that of Federer [1, 4.4.2, 5.1.6]. Consider a minimizing sequence T_i, viewed as locally integral currents in $M - B$ (cf. 9.1). By compactness, we may assume that the T_i converge to a minimizer T. The difficult part is

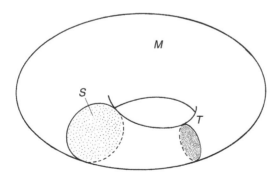

Figure 12.3.1. The area-minimizer T relatively homologous to S provides the least cross-sectional area.

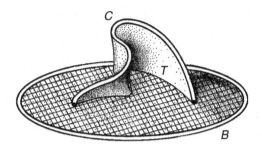

Figure 12.3.2. A soap film minimizing area in its relative homology class.

to show that T stays in the same relative homology class. Convergence (see 9.1) means that, for some Y and Z_1 in M with small mass,

$$E = T_i - T - Y - \partial Z_1$$

is supported in a small neighborhood of B. Let Y_1 be a minimizer in $\mathbf{R}^N$ with $\partial Y_1 = \partial Y$. Since ∂Y_1 lies in a small neighborhood of B and because $\mathbf{M}(Y_1) \leq \mathbf{M}(Y)$ is small, by monotonicity, 9.5, Y_1 lies in a small neighborhood of B. Let Z_2 be a minimizer in $\mathbf{R}^N$ with $\partial Z_2 = Y - Y_1$. Then $\mathbf{M}(Z_2)$ is small, and hence Z_2 lies in a small neighborhood of M. Let E_1 denote the projection of the cycle $E + Y_1$ onto B; then $E + Y_1 - E_1 = \partial Z_3$, with Z_3 in a small neighborhood of B. We now have

$$T_i - T = E_1 + \partial Z_1 + \partial Z_2 + \partial Z_3,$$

with E_1 in B and the Z_i in a small neighborhood of M. Projecting onto M yields

$$T_i - T = E_1 + \partial Z,$$

with Z in M, so that T lies in the same relative homology class as T_i, as desired.

Isoperimetric Inequality [Federer, 4.4.2] Very similar arguments and the Deformation Theorem 5.1 provide a constant γ, depending on M, such that, if X is an m-dimensional relative boundary in M, then there is a rectifiable current Y in M such that $X = \partial Y$ (mod B) and

(1) $$\mathbf{M}(Y)^{m/(m+1)} + M(\partial Y - X) \leq \gamma \mathbf{M}(X).$$

Moreover, every relative cycle with small mass is a relative boundary.

Remarks on Regularity Of course, away from B, a relatively homologically area-minimizing rectifiable current enjoys the same regularity as an absolutely area-minimizing rectifiable current. In addition, regularity results are known along the free boundary (see, e.g., Taylor [1], Grüter [1, 2], and Hildebrandt). In particular, along smooth (free) boundary, an area-minimizing hypersurface is a C^1 manifold with boundary except for a singular set of codimension at least 7 in the surface (Grüter [1, 2], which generalize easily from Euclidean space to manifolds). Incidentally, this regularity result also holds for unoriented hypersurfaces because they are locally orientable on the interior and along a free boundary, although not along a prescribed boundary.

Volume Constraints   The existence and regularity results of this section continue to hold for "isoperimetric" hypersurfaces satisfying volume constraints. On regularity, see §8.5 and Grüter [3].

12.4   Functions of Bounded Variation [Federer, 4.5.9; Giusti; Simon 3, §6; Ambrosio, Fusco, and Pallara, Chap. 3]   An important class of functions in analysis and the basis of the theory of sets of finite perimeter of De Giorgi [1, 4] is the space BV^{loc} of functions of locally bounded variation. A real-valued function f on $\mathbf{R}^1$ is in BV^{loc} if it agrees almost everywhere with a function g of finite total variation on any interval $[a, b]$,

$$\sup \left\{ \sum_{i=1}^{k} |g(x_i) - g(x_{i-1})| : k \in \mathbf{Z}, \, a \le x_0 \le \cdots \le x_k \le b \right\} < \infty,$$

or, equivalently, if the distribution derivative Df is a locally finite measure. Similarly, a real-valued function on $\mathbf{R}^n$ is in BV^{loc} if Df is a locally finite (vector-valued) measure. The associated space of currents $\{\mathbf{E}^n \llcorner f\}$ is precisely $\mathbf{N}_n^{loc}\mathbf{R}^n$, the locally normal currents of codimension 0. Here, we give a sampling from Federer's comprehensive theorem on BV^{loc}.

Theorem [Federer, 4.5.9]   *Suppose $f \in \mathbf{BV}^{loc}$.*
(1) If χ_s is the characteristic function of $\{x : f(x) \ge s\}$, then

$$Df = \int_{s \in \mathbf{R}} D\chi_s \, ds$$

and

$$|Df| = \int_{s \in \mathbf{R}} |D\chi_s| \, ds$$

almost everywhere.
(31) If $n \ge 1$, then there is a constant c such that

$$\|f - c\|_{L^{n/(n-1)}} \le n^{-1} \alpha_n^{-1/n} \int |Df|,$$

where α_n is the volume of the unit ball in $\mathbf{R}^n$. If f has compact support, then $c = 0$.

Remarks   The second statement of (13) just says that there is no cancellation in the first. In the notation of geometric measure theory, (13) becomes

$$\partial(\mathbf{E}^n \llcorner f) = \int \partial[\mathbf{E}^n \llcorner \{x : f(x) \geq s\}] \, d\,\mathscr{L}^1 s$$

and

$$\|\partial(\mathbf{E}^n \llcorner f)\| = \int \|\partial[E^n \llcorner \{x : f(x) \geq s\}\| \, d\mathscr{L}^1 s.$$

An excellent comprehensive treatment of BV appears in Giusti.

12.5   General Parametric Integrands [Federer 1, 5.1]

In many mathematical and physical problems the cost of a surface depends not only on its area but also on its position or tangent plane direction. The surface energy of a crystal depends on its orientation with respect to the underlying crystal lattice (cf. Taylor [2]). Therefore, one considers an integrand $\Phi(x, \xi)$ associating to a rectifiable current T a cost or energy

$$\Phi(T) = \int \Phi(x, \vec{T}(x)) \, d \, \|T\|.$$

The most important case remains the area integrand $A(x, \xi) = |\xi|$. One always requires Φ to be continuous and homogeneous in ξ. Usually, Φ is positive (for ξ nonzero), even ($\Phi(x, -\xi) = \Phi(x, \xi)$), and convex ($\Phi(x, \xi_1+\xi_2) \leq \Phi(x, \xi_1) + \Phi(x, \xi_2)$); that is, each

$$\Phi_a(\xi) \equiv \Phi(a, \xi)$$

is a norm. (If this norm is given by an inner product, then $\Phi(x, \xi)$ is a Riemannian metric.)

   The adjective *parametric* just means that $\Phi(T)$ depends only on the geometric surface T and not on its parameterization (geometric measure theory rarely uses parameterizations at all). Convexity of Φ means each Φ_a is convex,

(2) $$\Phi_a(\xi_1 + \xi_2) \leq \Phi_a(\xi_1) + \Phi_a(\xi_2),$$

with the geometric interpretation that the unit ball $\{\Phi_a(\xi) \leq 1\}$ is convex. It follows easily that planes are Φ_a minimizing; that is, if S is a portion of a plane and $\partial R = \partial S$, then

(3) $$\Phi_a(S) \leq \Phi_a(R).$$

When (2) holds, one says F is *semielliptic*.

Uniform convexity of Φ gives an estimate on the strength of the convexity inequality (1).

(4) $\Phi_a(\xi_1) + \Phi_a(\xi_2) - \Phi_a(\xi_1 + \xi_2) \geq c(|\xi_1| + |\xi_2| - |\xi_1 + \xi_2|),$

with the geometric interpretation for smooth Φ that the unit ball $\{\Phi_a(\xi) \leq 1\}$ has positive inward curvature. It follows easily that if S is a portion of a plane and $\partial R = \partial S$, then

(5) $\Phi_a(R) - \Phi_a(S) \geq \lambda(\mathbf{M}(R) - \mathbf{M}(S)).$

One says Φ is *elliptic*. Ellipticity, introduced by Almgren, seems to be the right hypothesis for theorems, but it is much more difficult to verify directly than uniform convexity. The two notions are equivalent in codimension 1.

All of these notions are invariant under diffeomorphisms of the ambient.

In general, semiellipticity implies lower semicontinuity and the existence of minimizers; ellipticity implies regularity of minimizers (see 8.5). Figure 12.5.1

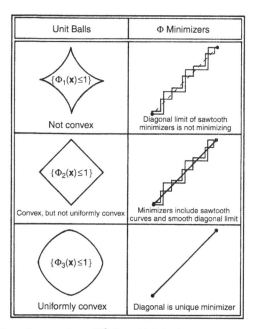

Figure 12.5.1. Three integrands on $\mathbf{R}^2$ for which horizontal and vertical directions are relatively cheap.

considers the cost of curves for three integrands $\Phi(\xi)$ on the plane for which horizontal and vertical directions are relatively cheap. Φ_1 is not convex and not lower semicontinuous, so limit arguments can fail to produce minimizers. Φ_2 is borderline convex and admits nonsmooth minimizers. Φ_3 is uniformly convex, and the cheapest path between two points is uniquely a straight line.

Soap Bubble Clusters

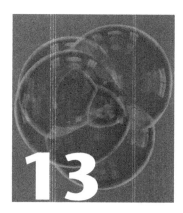

Soap bubble clusters and froths as in Figures 13.0.1 and 13.0.2 model biological cells, metallurgical structures, magnetic domains, liquid crystals, fire-extinguishing foams, bread, cushions, and many other materials and structures (see Weaire-Hutzler, Graner, Jiang *et al.*, and Zhang *et al.*). Despite the simplicity of the governing principle of energy or area minimization, the underlying mathematical theory is deep and still not understood.

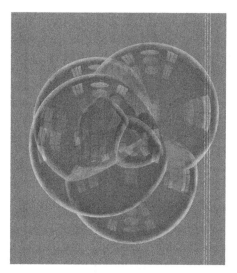

Figure 13.0.1. Bubble clusters seek the least-area way to enclose several volumes of air. Enhanced from "Computing Soap Films and Crystals," a video by the Minimal Surface Team, The Geometry Center. Computer graphics copyright John M. Sullivan; color version at www.math.uiuc.edu/~jms/Images.

Geometric Measure Theory. http://dx.doi.org/10.1016/B978-0-12-804489-6.00013-7

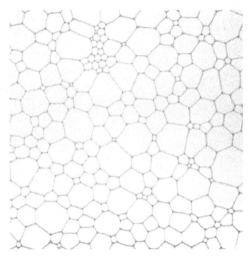

Figure 13.0.2. Soap bubble froth photographed by Olivier Lorderau at Rennes, used by permission.

A single soap bubble quickly finds the least-surface-area way to enclose the fixed volume of air trapped inside—the round sphere.

Similarly, bubble clusters seek the least-area way to enclose and separate several regions of prescribed volumes. This principle of area minimization alone, implemented on Ken Brakke's Surface Evolver (see 16.8), yields computer simulations of bubble clusters, as in Figure 13.0.1, from the video "Computing Soap Films and Crystals" by the Minimal Surface Team at the Geometry Center.

Do soap bubble clusters always find the absolute least-area shape? Not always. Figure 13.0.3 illustrates two clusters enclosing and separating the same five volumes. In the first, the tiny fifth volume is comfortably nestled deep in the crevice between the largest bubbles. In the second, the tiny fifth volume less comfortably sits between the medium-size bubbles. The first cluster has less surface area than the second. It might be still better to put the smallest bubble around in back.

As a matter of fact, until 2000 (see Chapter 14) it remained an open question whether the standard double bubble of Figure 13.0.4 is the least-area way to enclose two given volumes, as realized over the course of an undergraduate thesis by Foisy, despite a proof due to White that the solution must be a surface of revolution [Foisy, Theorem 3.4; cf. Morgan 3, Theorem 5.3]. It may seem difficult to imagine any other possibilities, until you see J. M. Sullivan's computer-generated competitor in Figure 13.0.5, which does, however, have more area and is unstable.

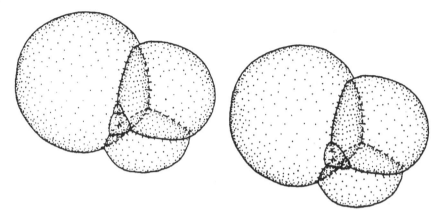

Figure 13.0.3. Soap bubble clusters are sometimes only relative minima for area. These two clusters enclose and separate the same five volumes, but the first has less surface area than the second.

Figure 13.0.4. A double bubble provides the least-area way to enclose two given volumes. Photo by Jeremy Ackerman, 1996.

More generally, both regions might have several components, wrapped around each other.

In general, it is a difficult open question whether each separate region is connected or whether it might conceivably help to subdivide the regions of prescribed volume, with perhaps half the volume nestled in one crevice here and the other half in another crevice there.

Similarly, it is an open question whether an area-minimizing cluster may incidentally trap inside "empty chambers," which do not contribute to the prescribed volumes. Figure 13.0.6 shows a 12-bubble with a dodecahedral empty chamber on

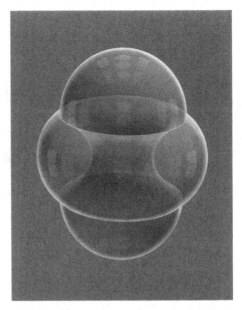

Figure 13.0.5. This nonstandard, computer-generated double bubble has more area and is unstable. Computer graphics copyright John M. Sullivan; color version at www.math.uiuc.edu/~jms/Images.

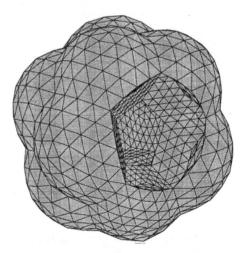

Figure 13.0.6. It is an open question whether area-minimizing clusters may have empty chambers, such as the dodecahedral chamber at the center of this 12-bubble. Graphics by Tyler Jarvis, Brigham Young University.

the inside, obtained by Tyler Jarvis of Brigham Young University using Brakke's Evolver. The computation postulated the empty chamber; without such a restriction, empty chambers probably never occur. Michael Hutchings proved that in a minimizing double bubble, there are no such empty chambers.

This chapter began with my AMS–MAA address in San Francisco, 1991, available on video [Morgan 4] and written up in Morgan [13, 12].

13.1 Planar Bubble Clusters Many of the fundamental questions remain open for planar bubble clusters—least-perimeter ways to enclose and separate regions of prescribe areas. For extensive progress, see Weaire-Hutzler and the work of Cox, Fortes, Graner, Heppes, Jiang, Teixeira, Vaz, and collaborators.

A proof of the planar double bubble conjecture appeared in 1993, the work of a group of undergraduates: Joel Foisy, Manuel Alfaro, Jeffrey Brock, Nickelous Hodges, and Jason Zimba (see Figure 13.1.1). Their work was featured in the 1994 AMS *What's Happening in the Mathematical Sciences*. A proof of the triple bubble conjecture by Wacharin Wichiramala appeared in 2002. Figure 13.1.2 shows computer calculations by Cox, Graner, *et al.* of the least-perimeter way to enclose and separate N regions of unit area for $3 \leq N \leq 42$. All cases for $N \geq 4$ remain unproven. Figure 13.1.3 shows four improvements Cox discovered later.

For large N, most regions are almost perfect regular hexagons, in line with Hales's Theorem 17.7 that regular hexagons provide a least-perimeter way to partition the whole plane into unit area. Cox, Graner, *et al.* noted and conjectured that the whole cluster approaches one big regular hexagon, as Fortes and Rosa

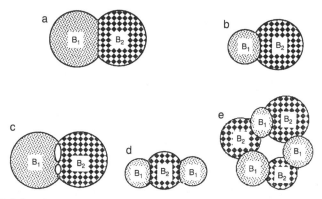

Figure 13.1.1. The standard planar double bubble (a and b), and not some exotic alternative with disconnected regions or empty chambers (c, d, and e), provides the least-perimeter way to enclose and separate two regions of prescribed area, as proved by a group of undergraduates [Foisy *et al.*, 1993].

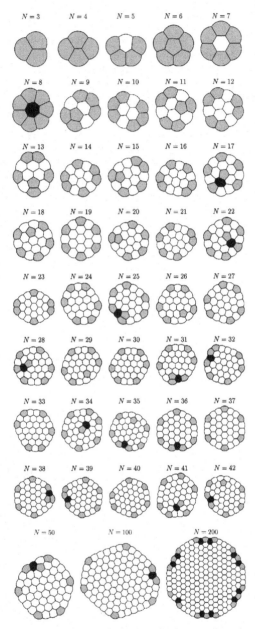

Figure 13.1.2. Computer calculations by Cox, Graner, *et al.* of the least-perimeter way to enclose *N* unit areas. A typical (white) region has six edges if on the interior and five edges if on the boundary. Regions with more edges are colored black; regions with fewer edges are shaded gray. Used by permission.

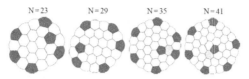

Figure 13.1.3. Cox later found improvements for four cases of Figure 13.1.2. Used by permission.

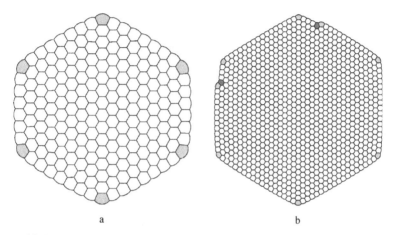

a   b

Figure 13.1.4. For a large number N of unit bubbles (here 200 and 1000), Cox *et al.* conjecture that the whole cluster will be approximately one big regular hexagon. The author, however, predicts that for N very large, the cluster will be approximately circular. Figures computed by Cox and Graner, used by permission.

proved is the best way to arrange small regular hexagons. Further evidence was provided by Cox and Graner by their best clusters of $N = 200$ and $N = 1000$ of Figure 13.1.4. On the other hand, I think that for very large N the whole cluster should become circular, to minimize the unshared outer perimeter, despite the cost of dislocations in the underlying hexagonal structure.

For formulas, it is often convenient to replace unit area with the area of a regular hexagon of unit edge length, perimeter 6, and area $A_0 = 3\sqrt{3}/2$. Cox *et al.* following Vaz–Fortes, give the following asymptotic estimate for the total interior and exterior perimeter p of a minimizing cluster of N bubbles of area A_0:

$$p \sim 3N + 3.10\sqrt{N},$$

based on the "perfect" 19-bubble. (Each little hexagonal region is billed for half of its six sides, with extra cost proportional to $\sqrt{N}$ for the unshared exterior perimeter.) This is probably fairly accurate if large clusters are roughly large hexagons.

Heppes and Morgan, assuming that very large clusters can become roughly circular with negligible additional internal cost, give the smaller estimate

$$p \sim 3N + \left(\pi^{3/2}/12^{1/4}\right)\sqrt{N} \approx 3N + 2.99\sqrt{N}.$$

The best proven upper bound has $\pi\sqrt{N}$; the best proven lower bound has only $\left(\sqrt{\pi A_0} - 1.5\right)\sqrt{N} \approx 1.36\sqrt{N}$.

13.1A Theorem [Heppes–Morgan, Theorem 3.1] *In the plane the least perimeter p for enclosing and separating N regions of area $A_0 = 3\sqrt{3}/2$ satisfies*

(1) $$3N + (\sqrt{\pi A_0} - 1.5)\sqrt{N} < p < 3N + \pi\sqrt{N} + 3.$$

Proof Sketch A poorer upper bound is given by nearly hexagonal subsets of the hexagonal honeycomb. The improvement here is obtained by rounding and stretching exterior boundary hexagons.

The much harder lower bound follows from keeping all terms in the last lines of Hales's deep proof (17.6) that $p > 3N$, or rather that for *unit* areas, $p/p_0 > N/2$, where p_0 is the perimeter of a regular hexagon of unit area. Indeed, as in 17.6, if s and t denote interior and exterior perimeter, then summing Hales's hexagonal isoperimetric inequality 17.4(1) yields

(2) $$(2s + t)/p_0 \geq N - 0.5 \sum a_i,$$

where a_i denotes how much more area is enclosed by an exterior edge than by a line segment, truncated so that $-1/2 \leq a_i \leq 1/2$. If $a_i \geq 0$ and the exterior edge has length t_i, then by comparison with a semicircle (Dido's inequality),

$$t_i \geq \sqrt{2\pi a_i} = a_i\sqrt{2\pi/a_i} \geq a_i\sqrt{4\pi}$$

because $a_i \leq 1/2$. Consequently, by (2),

$$2s + t \geq Np_0 - (0.5/\sqrt{4\pi})tp_0.$$

It follows that

$$p = s + t \geq Np_0/2 + t(0.5 - 0.25p_0/\sqrt{4\pi})$$
$$> Np_0/2 + \sqrt{4\pi N}(0.5 - 0.25p_0/\sqrt{4\pi}),$$

by the isoperimetric inequality. To rescale from areas 1 to areas $A_0 = 3\sqrt{3}/2$, multiply by $\sqrt{A_0} = 6/p_0$ to obtain

$$p > 3N + (\sqrt{\pi A_0} - 1.5)\sqrt{N}.$$

13.1B Connected Regions In the plane (but not in higher dimensions) the nontrivial general existence theory [Morgan 19] admits requiring regions to be connected, although they then must be allowed to bump up against each other.

13.2 Theory of Single Bubbles Simplikios's sixth-century commentary on Aristotle's *De Caelo* (see Knorr [p. 273], recommended to me by D. Fowler), referring incidentally to work no longer extant, reports that

> it has been proved . . . by Archimedes [287–212 BC] and Zenodorus [~200 BC] that of isoperimetric figures the more spacious one . . . among the solids [is] the sphere.

However, Archimedes and Zenodorus considered only a small class of solids, including, of course, the Platonic solids.

More than 2000 years later, H. A. Schwarz (1884) apparently gave the first complete proof, by a symmetrization argument. Schwarz symmetrization replaces slices by parallel hyperplanes in $\mathbf{R}^n$ with $(n-1)$-discs centered on an orthogonal axis. De Giorgi [2] (1958) gave a simple completion of an early argument of J. Steiner. Steiner symmetrization replaces slices by parallel lines with intervals centered on an orthogonal hyperplane. Perhaps the simplest known proof is based on the divergence theorem [Gromov, §2.1; Berger 1, 12.11.4]. This result also follows from the Brunn–Minkowski theorem [Federer, 3.2.41]. Furthermore, given known existence and regularity, there is a very simple symmetry proof (see Section 14.3 and [Hutchings, Corollary 2.8]). For more history, results, and references, see the excellent reviews by Burago and Zalgaller (especially §10.4) and Ros [1]. These isoperimetric results generalize to norms more general than area [Gromov, §2.1; Morgan 16, §10.6; Brothers and Morgan].

More General Ambients Round balls are known to be minimizing also in $\mathbf{S}^n$ and $\mathbf{H}^n$ [Schmidt]. Round balls about the origin are known to be minimizing in certain two-dimensional surfaces of revolution (see the survey by Howards *et al.*), in certain n-dimensional cones [Morgan and Ritoré], and in Schwarzschild-like spaces by Bray and Morgan, with applications to the Penrose inequality in general relativity. Morgan and Johnson [Theorem 2.2] show that in any smooth compact Riemannian manifold, minimizers for small volume are nearly round spheres. There are results on $\mathbf{R} \times \mathbf{H}^n$ by Hsiang and Hsiang; on $\mathbf{RP}^3$, $\mathbf{S}^1 \times \mathbf{R}^2$, and $\mathbf{T}^2 \times \mathbf{R}$ by Ritoré and Ros ([2]; [1], [Ritoré]); on $\mathbf{R} \times \mathbf{S}^n$ by Pedrosa; and on $\mathbf{S}^1 \times \mathbf{R}^n$, $\mathbf{S}^1 \times \mathbf{S}^n$, and $\mathbf{S}^1 \times \mathbf{H}^n$ by Pedrosa and Ritoré. In $\mathbf{RP}^3$, the following is the least-area way to enclose a given volume V: for small V, a round ball; for large V, its complement; and for middle-sized V, a solid torus centered on an equatorial $\mathbf{RP}^1$. In $\mathbf{RP}^n$ and $\mathbf{CP}^n$, minimizers are conjectured to be tubes about lower-dimensional $\mathbf{RP}^k$s and $\mathbf{CP}^k$s or complements [Berger 3, p. 318]. In particular, it remains an open conjecture that a minimizer in $\mathbf{CP}^2$ is a geodesic ball or complement. In the cubic flat three-torus, minimizers are conjectured to be balls, cylinders, slabs, or complements, whereas in the hexagonal flat three-torus, higher genus surfaces sometimes do better (see Ros [1, 2]). In Gauss space G^n ($\mathbf{R}^n$ with Gaussian density), minimizers are half spaces (see Chapter 15). In the sub-Riemannian–Heisenberg group, the minimizer is conjectured to be a particular surface of revolution generated by a geodesic (see Capogna *et al.*).

Methods of N. Smale showed that minimizers in Riemannian manifolds can have the small singular sets admitted by the regularity theory (Section 8.5). He proved, for example, that if you take a singular area-minimizing cone such as the one over $\mathbf{S}^3 \times \mathbf{S}^3$ in $\mathbf{R}^8$ (Section 10.7), cross it with a line, and intersect it with the unit sphere, you get a singular hypersurface in the sphere which is area minimizing in a tubular neighborhood for some metric on the sphere. If you make the metric huge on the rest of the sphere, it should be globally area minimizing for its volume.

13.3 Cluster Theory The existence of soap bubble clusters in $\mathbf{R}^n$ (see Figure 13.3.1) is guaranteed by the following theorem, proved in a more general context by Almgren [1, Theorem V1. 2] and specialized and simplified in Morgan [3, §4.4], where details can be found.

A *cluster* consists of disjoint regions $R_1, \ldots, R_m$ (n-dimensional locally integral currents of multiplicity 1) with volume $R_i = V_i$, complement R_0, and surface area

$$A = \frac{1}{2} \sum_{i=0}^{m} \mathbf{M}(\partial R_i).$$

Figure 13.3.1. There exists a "soap bubble cluster" providing the least-area way to enclose and separate m regions R_i of prescribed volumes. Photo by Jeremy Ackerman.

(By including R_0, the sum counts each surface twice, before multiplication by 1/2.) A region is not assumed to be connected. By 12.2, A is always less than or equal to the area of the union of the topological boundaries of the regions, with equality of course for nice regions.

13.4 Existence of Soap Bubble Clusters *In R^n, given volumes $V_1, \ldots, V_m > 0$, there is an area-minimizing cluster of bounded regions R_i of volume V_i.*

To outline a simple proof, we need a few lemmas. Lemma 13.5, an extremely useful observation of Almgren's (see Almgren [1, V1.2(3)] or Morgan [19, 2.2]), will let us virtually ignore the volume constraints in eliminating wild behavior.

13.5 Lemma *Given any cluster, there exists $C > 0$, such that arbitrary small volume adjustments may be accomplished inside various small balls at a cost*

$$|\Delta A| \leq C\,|\Delta V|$$

(see Figure 13.5.1).

Remark There is much freedom in the placement of the balls. Selecting two disjoint sets of such balls now will allow later proofs to adjust volumes at such locations, away from the main argument.

Proof Sketch By the Gauss–Green–De Giorgi–Federer theorem, 12.2, at almost all points of ∂R_i, R_i has a measure-theoretic exterior normal and the approximate tangent cone is a half-space. Hence, at almost every point of ∂R_1, for example, this half-space fits up against that of some other R_i, such as R_2. Gently pushing R_1 into R_2 costs $|\Delta A| \leq C_{12}\,|\Delta V|$. Combining over many such neighboring pairs yields an arbitrary small volume adjustment with $|\Delta A| \leq C\,|\Delta V|$.

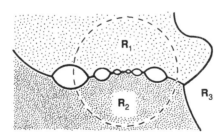

Figure 13.5.1. Gently pushing R_1 into R_2 at a typical border point yields a small volume adjustment with $|\Delta A| \leq C\,|\Delta V|$.

13.6 Lemma *An area-minimizing cluster is bounded in* $\mathbf{R}^n$.

Proof Let $V(r)$ denote the volume outside $\mathbf{B}(0, r)$; let $A(r)$ denote the area outside $\mathbf{B}(0, r)$. Truncation at almost any radius r saves $A(r)$, requires patching by the slice $\langle R_i, u, r+\rangle$ with $u(x) = |x|$ and

$$\mathbf{M}\langle R_i, u, r+\rangle \leq |V'(r)|$$

(4.11(3)), and requires replacing lost volume at a cost of $CV(r)$ for large r by Lemma 13.5. Therefore,

(1) $$|V'(r)| + CV(r) \geq A(r).$$

On the other hand, application of the isoperimetric inequality, 5.3, to the exterior of $\mathbf{B}(0, r)$ yields

(2) $$|V'(r)| + A(r) \geq \gamma V(r)^{(n-1)/n}.$$

Adding inequalities (1) and (2) yields

$$2|V'(r)| \geq -CV(r) + \gamma V(r)^{(n-1)/n} \geq \frac{1}{2}\gamma V(r)^{(n-1)/n}$$

for large r. If $V(r)$ is never 0,

$$n(V^{1/n})' = V^{-(n-1)/n} V' \leq -c < 0$$

for almost all large r, which contradicts positive and nonincreasing V. ∎

13.7 Sketch of Proof of Theorem 13.4 The main difficulty is that volume can disappear to infinity in the limit. First, we show that we can preserve some fraction of the volume. Let $\mathscr{C}_\alpha$ be a sequence of clusters with the prescribed volumes V_i with areas converging to the infimum A. We claim there are constants $S, \delta > 0$, such that if $\mathbf{R}^n$ is partitioned into cubes K_i of edge length S, then for some K_i,

(1) $$\text{vol}(R_{1,\alpha} \llcorner K_i) \geq \delta V_1^n.$$

Indeed, choose S large enough so that if $\text{vol}(R \llcorner K_i) \leq V_1^n$, then

$$\text{area}(\partial R \llcorner K_i) \geq \gamma(\text{vol}(R \llcorner K_i))^{(n-1)/n},$$

for some isoperimetric constant γ (12.3(1)). Let $\delta \leq \min\{(\gamma/A)^n, 1\}$. Then for each K_j,

$$\text{area}(\partial R_{1,\alpha} \llcorner K_j) \geq \frac{\gamma}{\max_i \text{vol}(R_{1,\alpha} \llcorner K_i)^{1/n}} \text{vol}(R_{1,\alpha} \llcorner K_j).$$

Summing over j yields

$$A \geq \frac{\gamma}{\max_i \text{vol}(R_{1,\alpha} \llcorner K_i)^{1/n}} V_1,$$

$$\max_i \text{vol}(R_{1,\alpha} \llcorner K_i) \geq (\gamma V_1/A)^n \geq \delta V_1^n,$$

proving the claim. Therefore, by translating the $\mathscr{C}_\alpha$, we may assume that for some fixed $r > 0$,

$$(2) \qquad\qquad \text{vol}(R_{1,\alpha} \llcorner \mathbf{B}(0, r)) \geq \delta V_1^n.$$

By a compactness argument (see Section 9.1), we now may assume that the $\mathscr{C}_\alpha$ converge to a limit cluster $\mathscr{C}$. By (2),

$$(3) \qquad\qquad \text{vol}(R_1) \geq \delta V_1^n.$$

The second step is a standard argument to show that $\mathscr{C}$ is area minimizing for its volume. Otherwise, a compact improvement of $\mathscr{C}$ could be used to improve substantially the $\mathscr{C}_\alpha$, with only a small volume distortion. The volume distortion could be corrected by truncation, slight homothetic contraction until no volume is too big, and the addition of tiny spheres. By Lemma 13.6, $\mathscr{C}$ is bounded.

Now if there was no volume loss to infinity, $\mathscr{C}$ solves our problem. If there was a volume loss, repeat the whole process with translations of the discarded material, obtained as the restriction of the $\mathscr{C}_\alpha$ to the exterior of an increasing sequence of balls. Countably many repetitions capture the total volume and yield a solution. (Since each is bounded, they do fit in $\mathbf{R}^n$.) Since the conglomerate solution must be bounded by Lemma 13.6, we conclude that only finitely many repetitions were actually needed.

General Ambient Manifolds   Theorem 13.4 holds in any smooth Riemannian manifold M with compact quotient M/Γ by the isometry group Γ.

Regularity results for minimizing clusters begin with the following.

13.8 Proposition *In a minimizing cluster, the rectifiable set* $S = \cup(\partial R_i)$ *is* $(\mathbf{M}, \varepsilon, \delta)$ *minimal:*

$$\mathcal{H}^{n-1}(S) \leq (1 + \varepsilon(r))\mathcal{H}^{n-1}(\varphi(S)),$$

where $\varepsilon(r) = 3Cr$, $r \leq \delta$.

Proof The proof depends on Lemma 13.5, which provides volume adjustments at cost $|\Delta A| \leq C|\Delta V|$. δ must be chosen small enough so that a δ-ball has small volume in the sense of Lemma 13.5 and so that any δ-ball is disjoint from a set of balls used to readjust volumes.

Consider a Lipschitz deformation inside an r-ball with $r \leq \delta$. The total volume distortion in moving S to $\varphi(S)$ is at most $r(\mathcal{H}^{n-1}(S) + \mathcal{H}^{n-1}(\varphi(S)))$ and certainly less than the volume of the r-ball. By Lemma 13.5, the volumes may be readjusted elsewhere at cost

$$|\Delta A| \leq C\,|\Delta V| \leq Cr(\mathcal{H}^{n-1}(S) + \mathcal{H}^{n-1}(\varphi(S)).$$

Since the original cluster is minimizing,

$$\mathcal{H}^{n-1}(S) \leq \mathcal{H}^{n-1}(\varphi(S)) + Cr(\mathcal{H}^{n-1}(S) + \mathcal{H}^{n-1}(\varphi(S)),$$

$$\mathcal{H}^{n-1}(S) \leq \frac{1 + Cr}{1 - Cr}\mathcal{H}^{n-1}(\varphi(S)) \leq (1 + 3Cr)\mathcal{H}^{n-1}(\varphi(S))$$

for $r \leq \delta$ small. Therefore, S is $(\mathbf{M}, \varepsilon, \delta)$ minimal for $\varepsilon(r) = 3Cr$.

Soap bubble clusters, as in Figures 13.0.1 and 13.3.1, consist of smooth surfaces meeting in threes at 120 degrees along curves, which in turn meet in fours at angles of arccos $(-1/3)$ or approximately 109 degrees at points, as recorded in the landmark experiments and analysis of the Belgian physicist J. Plateau (1873). In short order, the physicist E. Lamarle (1864) theoretically derived Plateau's laws for area minimization, assuming that soap clusters are piecewise smooth, consisting of smooth surfaces meeting along smooth curves meeting at isolated points. It was another century before Jean Taylor (1976) gave a rigorous mathematical proof without assuming the clusters piecewise smooth. Taylor used geometric measure theory and Fred Almgren's theory of $(\mathbf{M}, \varepsilon, \delta)$-minimal sets. A description of Taylor's work appears in a *Scientific American* article by Taylor and Almgren, who supervised her Ph.D. thesis. (See also Kanigel.) As another good result, Almgren and Taylor were married (see Figure 13.8.1). ∎

Figure 13.8.1. Jean Taylor and Fred Almgren at their wedding, with Rob and Ann Almgren. Photograph courtesy of the Almgren–Taylor family.

13.9   Regularity of Soap Bubble Clusters in R³ [Taylor, 4]. *A soap bubble cluster $\mathscr{C}$ in* **R**³ *((M, ε, δ)-minimal set) consists of real analytic constant-mean-curvature surfaces meeting smoothly in threes at 120° angles along smooth curves, in turn meeting in fours at angles of* $\cos^{-1}(-1/3) \approx 109°$.

Remark   The singular curves were proved $C^{1,\alpha}$ by Taylor [4]; C^{∞} by Nitsche [1]; and real analytic by Kinderlehrer, Nirenberg, and Spruck [Theorem 5.1].

Comments on Proof   Consider a linear approximation or tangent cone C at any singularity (which exists by monotonicity and further substantial arguments). By scaling, C is (**M**, ε, δ) minimal for $\varepsilon = 0$ and $\delta = \infty$. C must intersect the unit sphere in a "net" of geodesic curves meeting in threes at 120°, an extension of the more familiar fact that shortest networks meet only in threes at 120°. (This angle comes from a balancing condition for equilibrium. A junction of four curves could be profitably deformed to two junctions of three as in Figure 11.3.1.) In 1964, Heppes [1], unaware of the earlier work of Lamarle (1864), found all 10 such geodesic nets, shown in Figure 13.9.1. Taylor, adding a final case to complete the work of Lamarle, showed all but the first three cones to be unstable by exhibiting area-decreasing deformations as in Figure 13.9.2. In an ironic twist of fate, in the fourth case, the comparison surface provided by Taylor has more area than the cone; the correct comparison surface, which reduces area by pinching out a flat triangular surface in the center, had been given correctly by Lamarle.

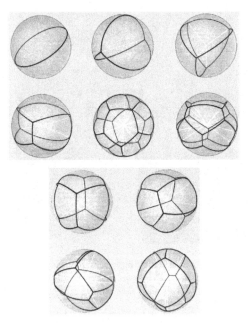

Figure 13.9.1. On the sphere there are exactly 10 nets of geodesics meeting in threes at 120°, providing 10 candidate cone models for soap bubble structures. Reprinted with permission from Almgren and Taylor. Copyright © 1976 by Scientific American, Inc. All rights reserved.

Actually back in 1964, in work not published until 1995, Heppes [2, Lemma 1] had already shown all but the first three cones to be unstable. When writing the paper, he checked *Math Reviews*, discovered Taylor's paper, and thus learned of Lamarle's work for the first time. Heppes finally met Taylor in 1995 at a special session on Soap Bubble Geometry in Burlington, Vermont, organized by the author.

Incidentally, an octahedral frame, whose cone, of course, cannot be in equilibrium since surfaces meet in fours, bounds at least five interesting soap films [Isenberg, color plate 4.6]. It is an open question whether the smallest of these five is, in fact, the minimizer.

Thus, for the approximating cone, there are just three possibilities, corresponding to a smooth surface or the two asserted types of singularities.

The really difficult part is to show that the cone is a good enough approximation to the original soap film. Taylor uses a deep method of Reifenberg [1–3], requiring the verification of a certain "epiperimetric inequality," which says roughly that cones near the special three are not too close to being minimizers themselves.

Of course, where the surface is regular, a classical variational argument yields constant mean curvature and hence real analyticity.

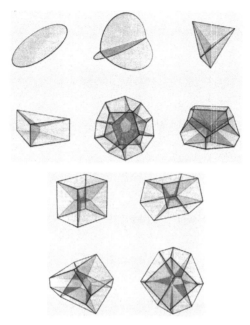

Figure 13.9.2. All but the first three cones are unstable, as demonstrated by the pictured deformations of less area. The fourth one actually should have a horizontal triangle instead of a vertical line segment in the center. Copyright © 1976 by Scientific American, Inc. All rights reserved.

Boundary Singularities   Lawlor and Morgan [1] (see also Morgan [18, Lecture I, Section 11.3, p. 91] and [Sullivan and Morgan, Problem 12]) describe 10 conjectured types of smooth boundary singularities of soap films, as in Figure 13.9.3. In studying boundary regularity, the definition of $(\mathbf{M}, 0, \delta)$-minimal set should be strengthened to allow sliding along the boundary, as suggested by Guy David.

13.10   Cluster Regularity in Higher Dimensions   Almgren [1, Theorem III.3(7)] proved that soap bubble clusters $((\mathbf{M}, \varepsilon, \delta)$-minimal sets) in $\mathbf{R}^n$ $(n \geq 3)$ are $C^{1,\alpha}$ almost everywhere. White [9] has proved that they consist of smooth constant-mean-curvature hypersurfaces meeting in threes at 120 degrees along smooth $(n-2)$-dimensional surfaces, which in turn meet in fours at equal angles along smooth $(n-3)$-dimensional surfaces, which meet in an $(n-4)$-dimensional set.

Brakke [2] has classified the polyhedral $(\mathbf{M}, 0, \infty)$-minimal cones in $\mathbf{R}^4$, which include the cone over the hypercube. It is conjectured that there are no non-polyhedral $(\mathbf{M}, 0, \infty)$-minimal cones below $\mathbf{R}^8$.

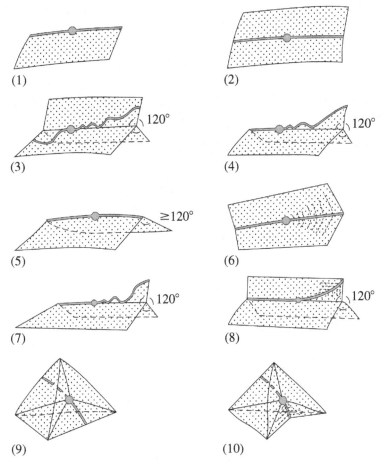

Figure 13.9.3. Ten conjectured types of soap film boundary singularities. Lawlor and Morgan [1, Figure 5.3].

For a simple treatment of 1-dimensional $(\mathbf{M}, \varepsilon, \delta)$-minimal sets, see Morgan [7]. (Almgren's regularity results technically do not apply [Almgren 1, IV.3(1), p. 96].) Taylor's [4] regularity results go through for soap bubble clusters in $\mathbf{R}^2$. Only in $\mathbf{R}^2$ does existence theory provide the alternative option of requiring that the regions be connected [Morgan 19]; such regions may bump up against each other.

13.11 Minimizing Surface and Curve Energies In some materials, not only interfacial surfaces but also singular curves may carry energetic costs. This modification alters behavior qualitatively as well as quantitatively, with four surfaces meeting along a singular curve, for example, as in Figure 13.11.1. Some

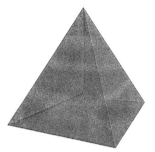

Figure 13.11.1. If singular curves carry energetic cost, the degree-four singularity of the soap film of Figure 11.3.1 can profitably decompose into two degree-three singularities, with four surfaces meeting along the singular curve between them. [Morgan and Taylor, Figure 3.]

results on existence, regularity, and structure appear in Morgan and Taylor and in Morgan [22, 3].

13.12 Bubble Cluster Equilibrium   For an equilibrium bubble cluster in $\mathbf{R}^n$, each region has a pressure p_i defined up to a constant (the external pressure p_{ext}) such that a normalized mean curvature of an interface between regions R_i and R_j is $p_i - p_j$. The rate of change of area with respect to volume $dA/dV_i = p_i - p_{\text{ext}}$. Equilibrium implies that

(1) $$\frac{n-1}{n} A = \sum (p_i - p_{\text{ext}}) V_i$$

[Cox *et al.* 1, Remark 4.4]. For an ideal gas $\sum p_i V_i = NT$ and hence

(2) $$\frac{n-1}{n} A = NT - p_{\text{ext}} V.$$

(Depending on units, there are usually some physical constants.) Condition (2) had an early if questionable derivation by S. Ross, with reference to an earlier suggestion by Tait.

Proof of (1)   Under scaling by a factor $(1 + t)$,

$$A(t) = A \cdot (1 + t)^{n-1}, \qquad V_i(t) = V_i \cdot (1 + t)^n,$$

$$(n-1)A = \frac{dA}{dt} = \sum (p_i - p_{\text{ext}}) \frac{dV_i}{dt} = n \sum (p_i - p_{\text{ext}}) V_i.$$

(Equivalently, the cluster is stationary for enthalpy $A - \sum (p_i - p_{\text{ext}}) V_i$, as observed by Graner, Jiang, *et al.* [note 42].)

13.13 Von Neumann's Law

In 2007, MacPherson and Srolovitz provided a generalization of Von Neumann's law for the growth rate of a bubble in a foam from two to higher dimensions. Their exact formula for the integral mean curvature is the culmination of decades of theoretical and numerical approximations by other workers.

One assumes that the foam is in equilibrium except for relatively slow diffusion of air and that the pressure differences are small compared to the absolute pressure (typically less than 0.1% for physical foams).

Von Neumann's law states that for a two-dimensional bubble with n edges, perhaps in a foam between plexiglass plates, the rate of change of area is proportional to $n - 6$:

$$(1) \qquad\qquad dA/dt = a(n - 6).$$

Derivation The diffusion of air through an interface is proportional to the pressure difference, which is in turn proportional to the mean curvature of the interface. Since pressure differences are small, the area is proportional to the amount of air, and dA/dt is proportional to the integral around the boundary of the curvature. By Gauss–Bonnet, the entire turning of 2π equals this integral plus n contributions of $\pi/3$ (corresponding to the interior angles of $2\pi/3$ by the equilibrium assumption). Therefore,

$$dA/dt = -a_1(2\pi - n\pi/3) = a(n - 6).$$

The MacPherson–Srolovitz generalization states that for a three-dimensional bubble with "mean width" L and total singular edge length E, the rate of change of volume satisfies

$$(2) \qquad\qquad dV/dt = a(E - 6L).$$

The mean width L is defined as the integral over all planes of the Euler characteristic of the intersection of the plane with the interior of the bubble. The Euler characteristic is just the number of pieces minus the number of holes in those pieces, typically 1 or 0. The integral can be computed by first integrating over all planes perpendicular to a given line through the origin and then averaging over all lines through the origin. For a sphere, this turns out to be the diameter times an unfortunate factor of 2. In fact, for any closed convex surface, it is twice the width averaged over all directions.

The MacPherson–Srolovitz formula (2) is provided by averaging Von Neumann's law (1) over intersections with all planes.

MacPherson and Srolovitz provide similar generalizations to all dimensions.

13.14 Helmholtz Free Energy

Without the physically valid assumption that the differences in pressure are a negligibly small fraction of the atmospheric pressure, it is the amount of air inside a bubble, and not the volume of air, which is fixed, along with the temperature T. Moreover, according to physics, it is not the surface potential energy U which is minimized but, rather, a certain combination of U and the entropy S of the enclosed air called the Helmholtz free energy:

$$F = U - TS.$$

Fortunately, the existence and regularity theory remains valid by the following.

Proposition   *A cluster which minimizes Helmholtz free energy for given amounts of air and fixed temperature exists and also minimizes surface energy for given volumes.*

Proof   For given volumes, we can minimize surface energy U by Theorem 13.4. Since entropy S depends only on the volumes, we have minimized Helmholtz free energy F as well. Conversely, any such minimizer of F is a minimizer of U. Since this minimum F is continuous in the given volumes and blows up as either volume goes to infinity (because U blows up) or to 0 (because S goes to minus infinity), we can now choose the given volumes to minimize F. This minimizer in particular minimizes U for given volumes, and the soap bubble regularity applies.   ∎

Exercises

13.1 Give an example of a smooth metric on the plane for which the least-perimeter way to enclose unit area is not connected.

13.2 Consider the plane with a cusp hill centered above each integer point on the x-axis on which any area $0 < A \leq 1$ can be enclosed with a curve of length A^2. What can you say about the least-perimeter enclosure of given area? (Note that the remark at the end of Section 13.7 on general ambient manifolds does not apply because the metric is not smooth.)

13.3 Prove that a length-minimizing network in the plane consists of lines meeting in threes at 120 degrees and possibly in twos at angles of at least 120 degrees at given boundary points. You may assume that the network consists of finitely many lines meeting at finitely many points.

Proof of Double Bubble Conjecture

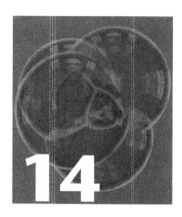

The year 2000 brought an announcement of the proof of the Double Bubble Conjecture in $\mathbf{R}^3$ by Hutchings, Morgan, Ritoré, and Ros. The Double Bubble Conjecture states that the "standard" familiar double soap bubble of Figure 14.0.1, consisting of three spherical caps meeting at 120 degrees, provides the least-perimeter way to enclose two prescribed volumes in $\mathbf{R}^3$ (or similarly in $\mathbf{R}^n$). Long believed (see Plateau [pp. 300–301] and Boys [p. 120]) but not published, the conjecture finally appeared in a 1991 undergraduate thesis at Williams College by Joel Foisy, who realized that no one knew how to prove it, despite the fact that a minimizer was known to exist and be a surface of revolution.

The 1990 Williams College NSF "SMALL" undergraduate research Geometry Group, including Foisy, proved the Double Bubble Conjecture in $\mathbf{R}^2$ using original geometric arguments and some computation.

In 1995, Joel Hass, Michael Hutchings, and Roger Schlafly announced a proof of the Double Bubble Conjecture for equal volumes in $\mathbf{R}^3$ using new theoretical tools and extensive computer computations. (See also Morgan [5], Hass and Schlafly, and Hutchings.)

In 2000, Hutchings, Morgan, Manuel Ritoré, and Antonio Ros announced a proof of the Double Bubble Conjecture for arbitrary volumes in $\mathbf{R}^3$ using a new instability argument. (See also Cipra.)

The Williams undergraduate research Geometry Group, consisting of Ben Reichardt, Cory Heilmann, Yuan Lai, and Anita Spielman, generalized the result from $\mathbf{R}^3$ to $\mathbf{R}^4$, and, for the case when the larger volume is more than twice the smaller, to $\mathbf{R}^n$.

Finally, in 2007, Reichardt proved the Double Bubble Conjecture for arbitrary volumes in $\mathbf{R}^n$, eliminating any need for component bounds by generalizing Proposition 14.18.

Geometric Measure Theory. http://dx.doi.org/10.1016/B978-0-12-804489-6.00014-9

143

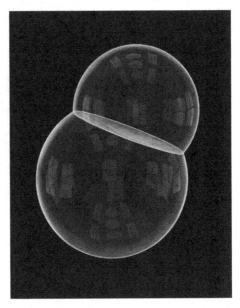

Figure 14.0.1. The standard double bubble provides the least-perimeter way to enclose and separate two prescribed volumes. Computer graphics copyright © John M. Sullivan; color version at www.math.uiuc.edu/~jms/Images/.

This chapter outlines the proof of the Double Bubble Conjecture in $\mathbf{R}^3$. A major difficulty is ruling out a bubble with three components, as in Figure 14.0.2. The main new idea is to prove such bubbles unstable by rotating pieces about well-chosen axes (see Proposition 14.15).

14.1 Proposition *For prescribed volumes v, w, there is a unique standard double bubble in $\mathbf{R}^n$ consisting of three spherical caps meeting at 120 degrees as in Figure 14.0.1.*

The mean curvature H_0 of the separating surface is the difference of the mean curvatures H_1, H_2 of the outer caps.

Proof Consider a unit sphere through the origin and a congruent or smaller sphere intersecting it at the origin (and elsewhere) at 120 degrees, as in Figure 14.1.1. There is a unique completion to a standard double bubble. Varying the size of the smaller sphere yields all volume ratios precisely once. Scaling yields all pairs of volumes precisely once.

The condition on the curvatures follows by the law of sines (see Figure 14.1.2) for $\mathbf{R}^2$ and hence for $\mathbf{R}^n$. (Plateau presented this figure for constructing standard double bubbles.) Since curvature is proportional to pressure difference, this condition implies that the pressure is well-defined (up to a constant). ∎

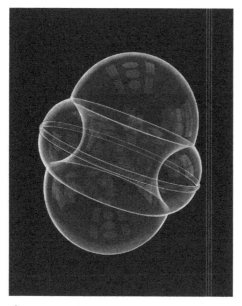

Figure 14.0.2. In $\mathbf{R}^3$, a major difficulty is ruling out a bubble with three components. Here, one region consists of a central bubble and a thin toroidal tube around the outside, whereas the second region consists of a larger toroidal tube in between. Computer graphics copyright © John Sullivan; color version at www.math.uiuc.edu/~jms/Images/.

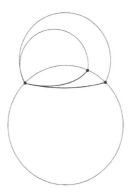

Figure 14.1.1. Varying the size of the smaller component yields all volume ratios precisely once.

14.2   Remark   Montesinos has proved the existence of a unique standard
k-bubble in $\mathbf{R}^n$ for $k \le n + 1$. (See also Sullivan and Morgan.)

The following symmetry theorem is based on an idea from Brian White, written up by Foisy [Theorem 3.4] and Hutchings [Theorem 2.6]. Incidentally, the

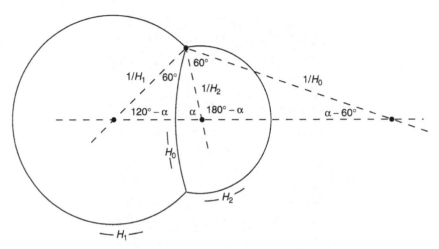

Figure 14.1.2. By the law of sines, the mean curvature H_0 of the separating surface is the difference of the mean curvatures H_1, H_2 of the outer caps.

same argument shows that an area-minimizing single bubble is a round sphere. The use of Steiner symmetrization across the axis hyperplanes to obtain central symmetrization dates back at least to Bieberbach (see the proof of the isodiametric or Bieberbach inequality on page 15).

14.3 Theorem *An area-minimizing double bubble in $\mathbf{R}^n$ is a hypersurface of revolution about a line.*

Proof Sketch (for details see the proof of Hutchings). In $\mathbf{R}^2$, the standard double bubble, which is symmetric about a line, is known to be the unique minimizer (Foisy *et al.*).

Consider a double bubble B in $\mathbf{R}^3$. We claim that some vertical plane splits both volumes in half. Certainly, for every $0 \leq \theta \leq \pi$, there is a vertical plane at angle θ to the xy-plane, which splits the first region in half. These planes can be chosen to vary continuously back to the original position, now with the larger part of the second region on the other side. Hence for some intermediate θ, the plane splits both volumes in half.

Now we may assume that the bubble B is symmetric under reflection across the plane. If not, reflect the half of less area across the plane to obtain a symmetric bubble of no more area. It must therefore have the same area and so actually either half would work. If each such half were part of a surface of revolution about a line in the plane, the original bubble B would have to be a surface of revolution, by Lemma 14.4. Thus, we may assume that B is symmetric under reflection across the plane.

Now turn everything to make our first vertical plane horizontal and obtain a second vertical plane of reflectional symmetry. Since the planes of reflectional symmetric are perpendicular, we may assume that they are the xz- and yz-planes. B is symmetric under the composition of the two reflections, i.e., under 180-degree rotation about the z-axis. Hence every vertical plane containing the z-axis splits both volumes in half.

We claim that at every regular point, the bubble B is orthogonal to the vertical plane. Otherwise, the smaller or equal half of B, together with its reflection, would be a minimizer with an illegal singularity (which could be smoothed to reduce area while maintaining volume). Now it follows that our bubble B in $\mathbf{R}^3$ is a surface of revolution.

Consider a double bubble B in $\mathbf{R}^n$ ($n \geq 3$). As before, we may assume reflectional symmetry across $n - 1$ orthogonal hyperplanes H_i and deduce symmetry under rotation about the intersection $H_i \cap H_j$ of any pair. It follows that B is a hypersurface of revolution about the line $\cap H_i$.

14.4   Lemma   *Suppose that both regions of a minimizing double bubble B in $\mathbf{R}^n$ ($n \geq 3$) are split in half by a hyperplane H. Let B_1, B_2 be the halves together with their reflections. If each B_i is symmetric about an $(n - 2)$-plane P_i in H, then they are both symmetric about P_1 (indeed, about $P_1 \cap P_2$).*

Proof Sketch   It suffices to show that at every regular point of B_2, the tangent plane contains the direction of rotation about $P_1 \cap P_2$. By symmetry, almost every point p of $B \cap H - P_1 - P_2$ is a regular point of B_1, B_2, and B (since a singular point yields a whole singular orbit). By analytic continuation, in a neighborhood of p, the tangent plane contains the directions of rotation about $P_1 \cap P_2$. By symmetry, this holds almost everywhere in B_2, except possibly at points where B_2 is locally contained in the orbit of $P_1 - P_2$. By the symmetry of B_2, such a piece of surface would continue down to H, where the symmetry of B_1 shows that as a piece of B it does not separate distinct regions, a contradiction.

14.5   Concavity (Hutchings Theorem 3.2)   *The least area $A(v, w)$ of a double bubble of volumes v, w in $\mathbf{R}^n$ is a strictly concave function. In particular, $A(v, w)$ is increasing in each variable.*

Proof Sketch   To illustrate the idea of the proof for the interesting case $n = 3$, we will just prove that for fixed w_0, $A(v, w_0)$ is nondecreasing, which suffices to prove Corollary 14.6. If not, then there is a local minimum at some v_0. For simplicity, we treat just the case of a strict local minimum. Consider a minimizing double bubble B of volumes v_0, w_0. By Theorem 14.3 and its proof, B is a surface of revolution about a line $L = P_1 \cap P_2$, where P_1 and P_2 are planes that divide both regions in half. Choose a plane P_3 near P_2 which divides the second region in half but does not contain L. We claim that it divides the first region in half. Otherwise, the half with smaller (or equal) area, reflected across the plane, would yield a bubble of no more area and slightly different volume, contradicting the assumption that v_0 is a strict local minimum. Therefore, P_3 splits both volumes in half. Now as in the proof of Theorem 14.3, B is symmetric about the line $L' = P_1 \cap P_3$ as well as about L. It follows that B consists of concentric spheres, which is impossible. Therefore, $A(v, w_0)$ must be nondecreasing as desired.

14.6 Corollary (Hutchings Theorem 3.4) *An area-minimizing double bubble in $\mathbf{R}^n$ has connected exterior.*

Proof If the exterior has a second, bounded, component, removing a surface to make it part of one of the two regions would reduce area and increase volume, in contradiction to $A(v, w)$ increasing. ∎

14.7 Corollary (Hutchings Theorem 3.5) *If the larger region of a minimizing double bubble has more than twice the volume of the smaller region, it is connected.*

Proof If not, create a new double bubble by calling the smallest component of the first region part of the second region. Of course, total volume $v' + w' = v + w$ does not change and area does not increase. Each region now has volume greater than w so that (v', w') is a convex combination of (v, w) and (w, v). By Concavity 14.5, area must increase, a contradiction. ∎

14.8 Decomposition Lemma (Cox *et al.* [2], Hutchings Lemma 4.1) *Consider a minimizing double bubble of volumes v, w in $\mathbf{R}^n$. If the first region has a component of volume $x > 0$, then*

$$2A(v, w) \geq A(v - x, w) + A(x) + A(v - x, w + x),$$

where $A(x) = A(x, 0)$ denotes the area of a sphere of volume x in $\mathbf{R}^n$.

Proof The result follows from the decomposition of Figure 14.8.1. Note that each piece of surface on the left appears twice on the right. ∎

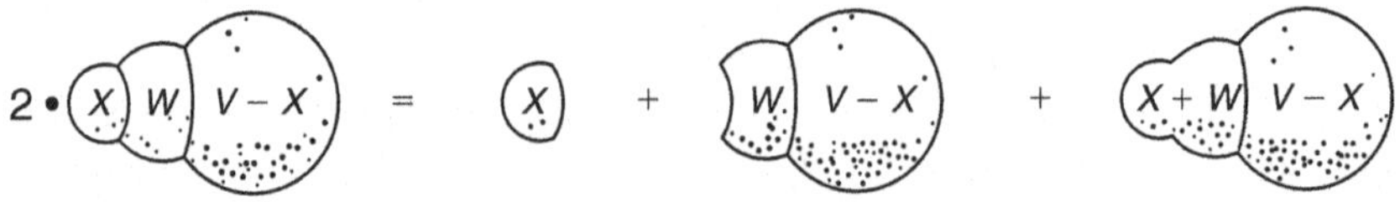

Figure 14.8.1. A useful decomposition.

14.9   Hutchings Basic Estimate (Hutchings Theorem 4.2)   *Consider a minimizing double bubble of volumes v, w in $\mathbf{R}^n$. If the first region has a component of volume $x > 0$, then*

(1) $$2A(v, \ w) \geq A(v + w) + A(w) + A(v)[v/x]^{1/n}.$$

In particular, a region has only finitely many components.

Proof   By Decomposition 14.8,

$$2A(v, \ w) \geq A(v - x, \ w) + A(x) + A(v - x, \ w + x).$$

Now by Concavity 14.5,

$$A(v - x, \ w) \geq \frac{v - x}{v} A(v, \ w) + \frac{x}{v} A(w),$$

Figure 14.9.1. Michael Hutchings obtained bounds on the number of components of an area-minimizing double bubble.

$$A(v-x,\ w+x) \geq \frac{v-x}{v}A(v,\ w) + \frac{x}{v}A(v+w).$$

Substituting and rearranging terms, using that $A(x) = A(v)\,[x/v]^{n-1/n}$, yields (1). Of course, this gives a bound on v/x and hence on the number of components of the first region. ∎

14.10   Hutchings Structure Theorem (Hutchings Theorem 5.1)

An area-minimizing double bubble in $\mathbf{R}^n$ is either the standard double bubble or another surface of revolution about some line, consisting of a topological sphere with a tree of annular bands (smoothly) attached, as in Figure 14.10.1. The two caps are pieces of round spheres, and the root of the tree has just one branch. The surfaces are all constant-mean-curvature surfaces of revolution, "Delaunay surfaces," meeting in threes at 120 degrees.

Proof Sketch Regularity, including the 120-degree angles, comes from applying planar regularity theory [Morgan 19] to the generating curves; also, the curves must intersect the axis perpendicularly. The bubble must be connected or moving components could create illegal singularities (or alternatively an asymmetric minimizer). By comparison with spheres centered on the axis and vertical

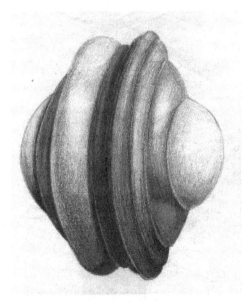

Figure 14.10.1. A nonstandard area-minimizing double bubble in $\mathbf{R}^n$ would have to consist of a central bubble with layers of toroidal bands. Drawing by Yuan Lai.

Figure 14.10.2. An intermediate piece of surface through the axis must branch into two spheres S_1, S_2.

hyperplanes, pieces of surface meeting the axis must be such spheres or hyperplanes. We claim that the number of such pieces is two (or three for the standard double bubble). If it were 1, an argument given by Foisy [Theorem 3.6] shows that the bubble could be improved by a volume-preserving contraction toward the axis $(r \to (r^{n-1} - \varepsilon)^{1/(n-1)})$. If it were 1, that piece of surface would not be separating any regions. If it were more than 2, some piece separates the two regions and eventually branches into two surfaces S_1 and S_2, as in Figure 14.10.2. We claim that S_1 and S_2 must be spherical. If, for example, S_1 were not spherical, replacing it (or part of it) by a spherical piece enclosing the same volume (possibly extending a different distance horizontally) would decrease area, as follows from the area-minimizing property of the sphere. Since everything else can be rolled around S_1 or S_2 without creating any illegal singularities, they must be spheres and the bubble must be the standard double bubble. The structure theorem now follows since the only possible structures are bubbles of one region in the boundary of the other. The same rolling argument implies that the root of the tree has just one branch.

The Hutchings basic estimate, 14.9, also has the following corollary.

14.11   Corollary   *Let $\tilde{A}(v, w)$ be the area of the standard double bubble in R^n of volumes v, w, or any other upper bound on the minimum double bubble area. Consider a minimizing double bubble of volumes v, $1 - v$. Then the first region has at most k components, where*

$$A(v)k^{1/n} = 2\tilde{A}(v, 1 - v) - A(1) - A(1 - v).$$

Proof   Let x be the volume of the smallest component of the first region so that the number of components is bounded by $k = v/x$. Since, of course, the minimum $A(v, w) \leq \tilde{A}(v, w)$, Corollary 14.11 follows immediately from 14.9(1).   ■

Remarks   Mathematica graphs of this function k (for $n = 2, 3, 4, 5$) appear in Figure 14.11.1. Some results are summarized in Table 14.11.1. In particular, in R^2 both regions are connected, from which the Double Bubble Conjecture in R^2 follows easily (as in Proposition 22 of Foisy *et al.*). The R^n bounds are elegantly deduced from the Hutchings basic estimate, 14.9, in Heilmann *et al.* [Proposition 5.3].

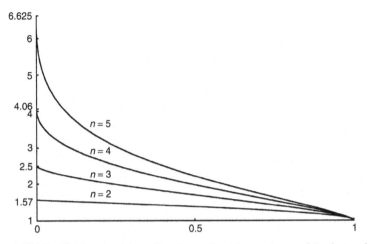

Figure 14.11.1. The function k bounding the number of components of the first region in a minimizing double bubble of volumes v, $1 - v$ in $\mathbf{R}^2$ through $\mathbf{R}^5$. B. Reichardt [Heilmann *et al.* Figure 2].

Table 14.11.1   Bounds on the number of components

	$\mathbf{R}^2$	$\mathbf{R}^3$	$\mathbf{R}^4$	$\mathbf{R}^5$	$\mathbf{R}^n$
Bounds on number of components in larger or equal region	1	1	1	2	3
Bounds on number of components in smaller region	1	2	4	6	2^n

14.12   Renormalization If in Corollary 14.11 we consider instead a minimizing bubble of volumes 1, w, then the bound k on the number of components of the first region satisfies

$$(2) \qquad A(1)k^{1/n} = 2\tilde{A}(1, w) - A(w + 1) - A(w).$$

14.13   Remark on Rigor There are a number of ways to prove rigorously the computational bounds of Table 14.11.1 apparent from the Mathematica plot of Figure 14.11.1. Unfortunately, obtaining useful bounds on the derivative of the plotted functions seems difficult as well as ugly. For $\mathbf{R}^3$, Hutchings *et al.* use a simple, weaker, *convex* bound to prove the larger region connected; of course, for a convex bound it is enough to check the endpoints. To prove that the smaller region has at most two components, they use a completely different auxiliary instability argument. For $\mathbf{R}^4$ (and incidentally $\mathbf{R}^3$), Heilmann *et al.* develop a piecewise convex bound to cover all volumes. Ultimately, Reichardt does not need any such bounds because he generalizes the main instability argument of Hutchings *et al.* to any number of components.

The following proposition, conjectured by Heilmann *et al.* [Conj. 4.10], provides one way to prove the bounds rigorously. It was proved by Marilyn Daily to win a $200 prize I offered in my Math Chat column (mathchat.org) on October 7, 1999.

14.14 Proposition [Daily] *In $\mathbf{R}^n$, let H_0, H_1, H_2, respectively, denote the mean curvature of a sphere of volume w, a sphere of volume $w + 1$, and the exterior of the second region of the standard double bubble of volumes 1, w, as suggested by Figure 14.14.1. Then*

$$2H_2 > H_0 + H_1.$$

Remark Since the derivatives of the terms of 14.12(1) are proportional to the mean curvatures (see Morgan [16]), Proposition 14.14 implies that the associated function k of 14.11 and 14.12 is increasing in w and hence decreasing in v.

Proposition 14.15 contains the main idea of the proof of the Double Bubble Conjecture in $\mathbf{R}^3$. The existence of points p_i where certain rotational eigenfunctions vanish implies instability (unless associated surfaces are all spheres and hyperplanes, which will be shown to be impossible). This idea goes back to the Courant nodal domain theorem, which relates the divisions of a domain by nodal sets to the position of the eigenvalues [Courant and Hilbert, VI-6, p. 452].

Consider a minimizing double bubble of revolution about the x-axis L in $\mathbf{R}^n (n \geq 3)$, with cross section Γ consisting of circular arcs $\overline{\Gamma}_0$ meeting the axis, and other arcs $\overline{\Gamma}_i$ meeting in threes, with interiors Γ_i (see Figures 14.16.1, 14.17.1). Consider the map f : $\Gamma - L \rightarrow L \cup \{\infty\}$, which maps $p \in \Gamma$ to the point $L(p) \cap L$,

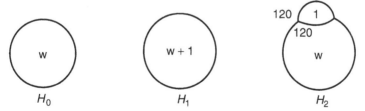

Figure 14.14.1. It is conjectured that the pictured curvatures satisfy $H_2 > (H_0 + H_1)/2$. This conjecture would provide an elegant way to prove rigorously the bounds of Table 14.11.1.

where $L(p)$ denotes the normal line to Γ at p. If $L(p)$ does not meet L, we define the image of p as $f(p) = \infty$.

14.15 Proposition (Hutchings *et al.* Proposition 5.1) *Consider a minimizing double bubble of revolution about the x-axis L in* $\mathbf{R}^n(n \geq 3)$. *Suppose that there is a minimal set of points* $\{p_1, \ldots, p_k\}$ *in* $\cup \Gamma_i$ *with* $x = f(p_1) = \cdots = f(p_k)$ *which separates* Γ. *(See Figure 14.15.1.)*

Then every component of the regular set which contains some p_i *is part of a sphere centered at x (if x* $\in$ *L) or part of a hyperplane orthogonal to L (in the case x* $= \infty$).

Proof Sketch for $\mathbf{R}^3$. Consider rotation about the line perpendicular to the cross-sectional plane at x. The normal component u of the rotation vector field vanishes on the circular orbits of the p_i about L. Let u_1 and u_2 denote the restrictions of u to the left and right portions of the cluster. Since the bubble is assumed to minimize area, both variations have nonnegative second variation of area. But their sum, a rotation vector field, has second variation 0. Therefore, u_1 has second variation 0 and is an eigenfunction. Since u_1 vanishes on the right portion of the cluster, by unique continuation for eigenfunctions $u_1 = 0$. Similarly, $u_2 = 0$. The proposition follows.

Actually, the proof in Hutchings *et al.* divides the cluster into four portions as in Figure 14.15.1. In October 2002, W. Wichiramala pointed out to me the simpler argument with two portions, new to this edition.

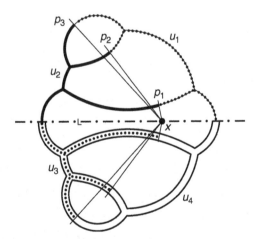

Figure 14.15.1. Rotation about an axis through x, which is tangential at the separating set p_1, p_2, p_3, leads to a proof of instability.

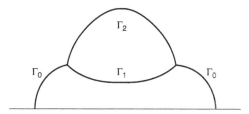

Figure 14.16.1. There is no nonstandard minimizing double bubble of revolution with connected regions. [Hutchings *et al.*]

14.16 Corollary *There is no nonstandard minimizing double bubble in* $\mathbf{R}^n$ *in which both regions and the exterior are connected, as in Figure 14.16.1.*

Proof The line equidistant from the two vertices intersects the axis L in a point p (unless the line is horizontal, which we will consider next). Hence, Γ_1 and Γ_2 each have an interior point farthest from or closest to p so that $p \in f(\Gamma_1) \cap f(\Gamma_2)$. By Proposition 14.15, Γ_1 and Γ_2 are both spherical, which is impossible.

If the line is horizontal, Γ_1 and Γ_2 each have an interior point farthest left or right so that $\infty \in f(\Gamma_1) \cap f(\Gamma_2)$. By Proposition 14.15, Γ_1 and Γ_2 are both vertical, which is impossible. ∎

14.17 Corollary *Consider a minimizing double bubble in* $\mathbf{R}^n$ *of three components and connected exterior as in Figure 14.17.1. Then there is no* $x \in \mathbf{R}$ *such that* $f^{-1}(x) - \Gamma_0$ *contains points in the interiors of distinct* Γ_j *which separate* Γ.

Proof If so, there must be points in Γ_1, Γ_2, or Γ_3. By Proposition 14.15, one of them is spherical. By so-called "force balancing" (Korevaar, Kusner, and Solomon), when two spherical pieces meet, the third piece is also spherical. Therefore, Γ_1, Γ_2, and Γ_3 are all spherical. But since Γ_2 bounds a region of positive pressure by Concavity 14.5 and hence Γ_2 curves to the right from its vertex with Γ_1, Γ_1 and Γ_2 cannot both be spherical. ∎

14.18 Proposition (Hutchings *et al.* Proposition 5.8) *There is no minimizing double bubble in* $\mathbf{R}^n$ *in which the region of smaller or equal pressure is connected, the other region has two components, and the exterior is connected, as in Figure 14.17.1.*

Remarks on Proof Corollary 14.17 reduces this to plane geometry, aided by a few simple facts about surfaces of Delaunay. Finding the separating set of points requires consideration of a number of cases, as suggested by Figure 14.18.1.

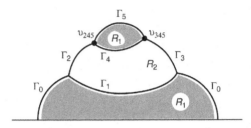

Figure 14.17.1. A candidate double bubble with three components. [Hutchings *et al.*]

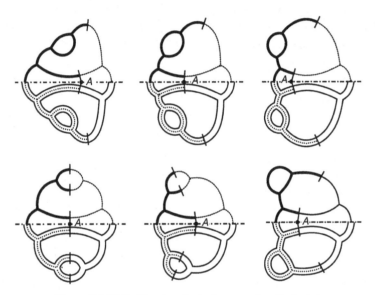

Figure 14.18.1. The six principal cases to be eliminated.

14.19 Proposition *In a minimizing double bubble in* $\mathbf{R}^n$, *the smaller region has larger pressure.*

Proof Consider the function $A(v, 1-v)$ giving the least area enclosing and separating regions of volume $v, 1-v$. By Concavity 14.5, A is strictly concave and, of course, symmetric about $v = 1/2$. Since one way to obtain nearby $(v, 1-v)$ is to vary the separating surface of mean curvature such as H, with $dA/dv = 2H$, the left and right derivatives of A must satisfy

$$A'_R \leq 2H \leq A'_L.$$

Consequently, H is positive for $v < 1/2$ and negative for $v > 1/2$. In other words, the smaller region has larger pressure. ∎

14.20   Theorem (Hutchings *et al.* **Theorem 7.1)**   *The standard double bubble in* $\mathbf{R}^3$ *is the unique area-minimizing double bubble for prescribed volumes.*

Proof   Let B be an area-minimizing double bubble. By Corollary 14.11 and Proposition 14.19, either both regions are connected or one of larger volume and smaller pressure is connected and the other of smaller volume and larger pressure has two components. By the Hutchings structure theorem, 14.10, B is either as in Figure 14.16.1 or as in Figure 14.17.1. By 14.16 and 14.18, B must be the standard double bubble. ∎

Remark   Although the final competitors are proved unstable, earlier steps such as symmetry (14.3) assume area minimization. It remains conjectural whether the standard double bubble is the unique *stable* double bubble.

14.21   Open Questions   It is conjectured by Hutchings *et al.* that the standard double bubble in $\mathbf{R}^n$ is the unique stable double bubble. Sullivan [Sullivan and Morgan, Proposition 2] has conjectured that the standard k-bubble in $\mathbf{R}^n$ ($k \leq n + 1$) is the unique minimizer enclosing k regions of prescribed volume. For now, even the standard triple bubble in $\mathbf{R}^3$ (Figure 13.3.1) seems inaccessible.

14.22   Physical Stability   As explained in Section 13.14, the technically correct physical soap cluster problem is to minimize the Helmholtz free energy $F = U - TS$ to enclose and separate given quantities rather than volumes of gas (at fixed temperature T), although the difference is negligible in practice. Here U is surface energy and S is entropy of the enclosed gas. To show that every round sphere minimizes F for a single given quantity of gas, since a round sphere minimizes surface area and hence U for fixed volume, it suffices to show that the number N of gas moles is an increasing function of volume v, which holds by scaling if for example N is proportional to Pv^n for $n > 1/3$ (for an ideal gas $n = 1$). Here is the similar result for double bubbles in more detail:

Proposition   *Assuming that the number N of gas moles is proportional to Pv^n for $n > 1/3$, a standard double bubble minimizes Helmholtz free energy for enclosing and separating two given masses of air.*

Proof   The Helmholtz free energy of a double bubble depends on the surface area and the volumes. For given volumes, the standard double bubble minimizes surface energy (Theorem 14.20) and hence F. As either volume goes to 0, the entropy S goes to minus infinity and F goes to infinity. As either volume goes to infinity, the surface area and hence F go to infinity. Therefore for given masses of air, some standard double bubble minimizes F. To show that every standard double bubble minimizes F for given masses, it suffices to show that the numbers of moles are increasing functions of the volumes. Since P — external pressure is proportional to curvature κ, it suffices to show that κv^n is an increasing function of volume. Since $n > 1/3$, by a scaling argument it suffices to prove this as one volume increases and the other remains fixed. The volume is an increasing fraction of the volume w of a sphere of curvature κ. Since by scaling κw^n is increasing, so is κv^n. ■

Exercises

14.1 Prove the following version of the Decomposition Lemma 14.8 [Hutchings p. 300]:

$$2A(v, w) \geq A(x) + A(w) + A(v - x) + A(v + w).$$

14.2 Prove the Double Bubble Conjecture in $\mathbf{R}^2$ as follows. You may assume that solutions exist to such problems and consist of circular arcs meeting only in threes at 120 degrees. The whole solution must be connected, or moving one piece against another—still a solution—would contradict regularity. The problem is that each region may have several components and that there may be empty space on the inside, not part of either region of prescribed area.

  a. Given areas A_1, A_2, consider a double bubble B which is the least-perimeter way to enclose and separate regions of area at least A_1 and A_2. Explain why the exterior of B is connected—that is, why B has no empty space on the inside.

  b. B must have an outermost component (see Figure 2.4.1 from Foisy below). Explain why sliding it around the circle it is on cannot bump into anything else, and conclude that B is a standard double bubble.

  c. Since the perimeter $P_0(A_1, A_2)$ of a standard double bubble is increasing in A_1, A_2, argue that B actually has areas A_1, A_2 and finish the proof.

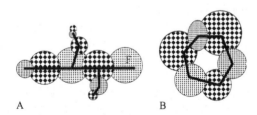

A B

[Foisy, Figure 2.4.1] Since the exterior is assumed to be connected as in A, the associated graph has an endpoint in a component with two edges and two vertices. If the exterior were disconnected, then a cycle as in B could result.

The Hexagonal Honeycomb and Kelvin Conjectures

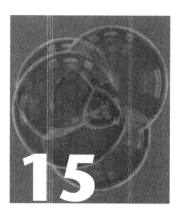

The Hexagonal Honeycomb Conjecture states that regular hexagons as in Figure 15.0.1 provide the most efficient (least-perimeter) way to divide the plane into unit areas. A proof was announced by Thomas Hales of the University of Michigan in 1999, the same Hales who recently proved the 1611 Kepler sphere-packing conjecture (see also Peterson [2], Klarreich, and *Notices AMS* 47(2000), pp. 440–449).

The Kelvin Conjecture describes a candidate for the least-perimeter way to divide three-space into unit volumes. A counterexample was announced by Denis Weaire and Robert Phelan of Trinity College, Dublin, in 1994 (see also Peterson [1] and Klarreich). Whether their counterexample is optimal remains open.

In 2008 Morgan [28] provided a short proof of the existence of a least-perimeter partition of $\mathbf{R}^n$ into unit volumes.

15.1 Hexagonal Honeycomb History

Since antiquity philosophers and honeybees have regarded hexagons as the ideal way to partition the plane into equal areas, as in the honeycomb of Figures 15.1.1 and 15.1.2. Around 36 BC,

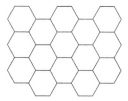

Figure 15.0.1. The Hexagonal Honeycomb Conjecture, proved by Hales in 1999, states that regular hexagons provide the most efficient (least-perimeter) way to divide the plane into unit areas [Hales].

Geometric Measure Theory. http://dx.doi.org/10.1016/B978-0-12-804489-6.00015-0

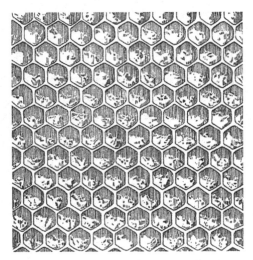

Figure 15.1.1. The bees' honeycomb illustrates the efficiency of using hexagons to enclose equal spaces with the least partitioning. From T. Rayment, *A cluster of bees* (The Bulletin, Sydney), as it appeared in D'Arcy Thompson's *On Growth and Form*, p. 109.

Figure 15.1.2. Bees at work. Photo by I. Kitrosser (Réalités 1 (1950), Paris), as it appeared in Herman Weyl's classic book, *Symmetry* [p. 84].

Figure 15.1.3. St. John's Abbey—University Church, St. John's University, my favorite place to speak on the hexagonal honeycomb.

Marcus Terentius Varro, in his book *On Agriculture* [III, xvi.5], wrote about the bees' honeycomb, "The geometricians prove that this hexagon...encloses the greatest amount of space." Zenodorus (∼200 BC) had proved the regular hexagon superior to any other hexagon, triangle, or parallelogram [Heath, pp. 206–212].

The next major advance came in 1953, when L. Fejes Tóth ([2, Chapter III, §9, p. 84] or [3, Corollary Section 26, p. 183] after [4]; see also [1] or [3, Section 29, pp. 206–208] on "wet films") used a simple convexity argument to prove regular hexagons superior in the very restricted category of convex (or polygonal) regions, modulo a truncation argument as in our Proposition 15.3. Many mathematicians came to have the erroneous impression that the problem was solved (see [Weyl, p. 85]).

Finally in 1999, Hales announced a proof in the category of connected regions.

When I arrived at St. John's University to give a talk on this topic, I was greeted by the church of Figure 15.1.3.

A difficulty with the honeycomb and its symmetry was pointed out in 1994 by Gary Larson (Figure 15.1.4).

In 2014 Caroccia and Maggi proved a quantitative version on the torus.

15.2   Definition of Clusters in $\mathbf{R}^2$ We will define a cluster C in $\mathbf{R}^2$ as a smooth, locally finite graph, with each face included in a unique region (nonempty union of faces) R_i or the exterior. We will usually assume that each R_i is connected (consists of a single face). For infinite clusters, we consider the perimeter $P(r)$ and

"Face it, Fred — you're lost!"

Figure 15.1.4. Gary Larson pointed out another difficulty with the honeycomb and its symmetry. THE FAR SIDE © 1994 FarWorks Inc. used by permission. All rights reserved.

area $A(r)$ inside the ball $\mathbf{B}(0, r)$. The *truncated cluster* $C_0(r)$ consists of the $n(r)$ regions completely contained inside the ball $\mathbf{B}(0, r)$, with perimeter $P_0(r)$ and area $A_0(r)$.

15.3 The Truncation Lemma (Morgan [8, proof of Theorem 2.1]) *Let C be a cluster of connected regions of area at most 1 in $\mathbf{R}^2$. Then*

$$\liminf_{r \to \infty} \frac{P_0(r)}{A_0(r)} \leq \limsup_{r \to \infty} \frac{P(r)}{A(r)} = \rho.$$

Proof For almost all r, $P'(r)$ exists and bounds the number of points where cluster boundaries meet the circle $\mathbf{S}(0, r)$, by the coarea formula 3.13 applied to the function $f(w) = r$. Given $\varepsilon > 0$, $r_0 > 0$, we can choose $r \geq r_0$ such that $P'(r) < \varepsilon A(r)$. Otherwise, for almost all $r \geq r_1 \geq r_0$,

$$P' \geq \varepsilon A > \frac{\varepsilon}{2\rho} P.$$

Since P is nondecreasing, for large r,

$$P(r) \geq P(r_1)e^{(\varepsilon/2\rho)(r-r_1)} > 2\rho\pi r^2,$$

which implies that $A(r) > \pi r^2$, a contradiction.

Therefore, in forming the truncated cluster C_0 from the restriction of C to $\mathbf{B}(0, r)$, at most $\varepsilon A(r)$ regions are discarded, and

$$\frac{P_0(r)}{A_0(r)} \leq \frac{P(r)}{(1 - \varepsilon)A(r)} \leq \frac{\rho}{1 - \varepsilon},$$

as desired.   ∎

The following inequality is the central idea of Hales's proof. Its penalty terms for bulging outward and for using more than six edges make the hexagon superior to the circle.

15.4   Hexagonal Isoperimetric Inequality (Hales Theorem 4)

Consider a curvilinear planar polygon of N edges, area A least $3.6/N^2$, and perimeter P. Let P_0 denote the perimeter of a regular hexagon of area 1. For each edge, let a_i denote how much more area is enclosed than by a straight line, truncated so that $-1/2 \leq a_i \leq 1/2$. Then

(1) $$P/P_0 \geq \min\{A, 1\} - .5 \sum a_i - c(N - 6),$$

with, for example, $c = 0.0505/2\sqrt[4]{12} \approx 0.013$, with equality only for the regular hexagon of unit area.

Remark   The truncation in the definition of a_i is necessary to prevent counter-examples as in Figure 15.4.1.

Proof Sketch   It is convenient to work in the larger category of *immersed* curvilinear polygons, counting area with multiplicity and sign. Then one may assume that all the edges are circular arcs (or straight lines). Without the truncation condition, one could further assume that all the edges have the same curvature since moving to that condition at constant net area decreases perimeter. With the truncation condition, the reductions are a little more complicated. Let x_i denote the excess area enclosed by an edge (in comparison with a straight line, before truncation, so that,

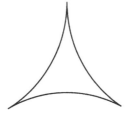

Figure 15.4.1. Without truncation in the definition of the a_i, this would be a counter-example to (1) [Hales].

for example, $|a_i| = \min\{|x_i|, 1/2\}$). We may assume that all $x_i \in (-1/2, 1/2)$ have the same curvature, and that all $x_i \in [-1/2, 1/2]$ have the same sign. We may assume that each $x_i \geq -1/2$, since increasing it to $-1/2$ decreases perimeter, leaves truncated area a_i at $-1/2$, and increases total area. Hales now treats separately two cases:

CASE 1   some $x_i > 1/2$,

CASE 2   every $|x_i| \leq 1/2$.

The proofs use five lower bounds on perimeter:

(L_N) The perimeter of the N-gon determined by the N vertices is at least as great as the perimeter L_N of the regular N-gon of the same area.

(L_+) "Standard isoperimetric inequality." The perimeter is at least the perimeter L_+ of a circle of area A, namely $2\sqrt{\pi A}$.

(L_-) The figure obtained by reflecting each edge across its chord has at least as much perimeter as a circle of the same area.

(L_D) "Dido's inequality." Each edge is at least as long as a semicircle enclosing the same area x_i.

(L'_N) If all the chords have length at most 1 and $|\Sigma x_i| \leq \pi/8$, then the perimeter is at least L_N times the length of a circular arc of chord 1 enclosing area $|\Sigma x_i| / L_N$. For example, when $\Sigma x_i = 0$, $L_{N'} = L_N$.

This last estimate is new, applying L_N to the original figure. It follows immediately from the chordal isoperimetric inequality, 15.5, below.

Before treating Cases 1 and 2, Hales treats the easy case of digons ($N = 2$), which uses nothing more than the standard isoperimetric inequality L_+.

Case 1 also is easy, using L_+, L_D, and a construction that reflects the edges enclosing negative area.

Case 2 is more delicate, using L_+ for Σx_i very positive, L_- or L_D for Σx_i very negative, and L_N in between. For $N \geq 100$, L_+ and L_D alone suffice. The subcases $N = 6$ or $N = 7$, $\Sigma x_i \approx 0$, require special attention and L'_N or the chordal isoperimetric inequality.

15.5   Chordal Isoperimetric Inequality (Hales Proposition 6.1)

Consider an immersed curvilinear polygon as in Figure 15.5.1 of length L, net excess area X over the chordal polygon, $|X| \leq \pi/4$, and each chord length at most 1. Let L_0 be the length of the chordal polygon. Then the length L is at least L_0 times the length arc($|X|/L_0$) of a circular arc of chord 1 enclosing area $|X|/L_0$.

The function L_0arc($|X|/L_0$) is increasing in L_0.

Remark   The constant $\pi/4$ is sharp, as may be seen by taking the curvilinear polygon to be a circle of area $X > \pi/4$ and the chordal polygon to be a unit chord transversed back and forth. By the standard isoperimetric inequality, the

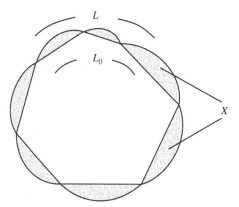

Figure 15.5.1. The chordal isoperimetric inequality estimates the perimeter L of a curvilinear polygon in terms of the perimeter L_0 of the chordal polygon and the excess shaded area X.

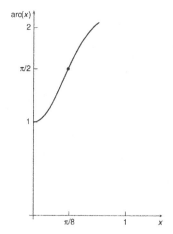

Figure 15.5.2. The length $\mathrm{arc}(x)$ of a circular arc enclosing area x with a unit chord is convex up to a semicircle ($x = \pi/8$) and concave thereafter. It follows that $\mathrm{arc}(x)/x$ is decreasing.

circle has less perimeter than two congruent arcs enclosing area $X/2$, i.e., $L < L_0\mathrm{arc}(X/L_0)$. We thank our colleague Steven Miller for help in finding the sharp constant.

Proof (A simplification and strengthening of Hales's original proof). We may assume that each edge is a circular arc. Since the derivative of $\mathrm{arc}(x)$ is just the curvature of the arc, $\mathrm{arc}(x)$ is convex up to a semicircle with $x = \pi/8$, as in Figure 15.5.2. Once we reduce to arcs at most semicircles, we will deduce the lemma by convexity.

We may assume that each excess area x_i has the same sign, by moving two with opposite signs closer to 0, without changing X, reducing length, until one is 0. We may assume that $X \geq 0$.

We now consider the less constrained problem of enclosing area X by arcs above chords on the x-axis of length at most 1 and total chord length L_0, with longest chord no longer than the sum of the lengths of the others. The minimizer consists of circular arcs. To be in equilibrium, they must have the same curvature because curvature is the rate of change of length with respect to area.

It pays to lengthen chords of (equal or) higher arcs since moving a tiny subarc from the middle of a lower arc to the middle of a higher arc would leave L_0 fixed and increase X. Now reducing the area under the lower arc to restore X reduces length. (This simplifying argument was suggested by Stewart Johnson.)

Hence, if $L_0 \leq 1$, a minimizer is a single arc on chord L_0, and its length

$$L_0 \text{arc}(|X|/L_0^2) \geq L_0 \text{arc}(|X|/L_0),$$

as desired.

If $1 < L_0 \leq 2$, by hypothesis the length of the longest chord is at most $L_0/2$ and a minimizer consists of arcs of equal curvature over two chords of length $L_1 = L_2 = L_0/2$. If one of the arcs were more than a semicircle, X would exceed the area of a circle of unit diameter, $\pi/4$, a contradiction. Similarly if $L_0 > 2$, a minimizer consists of $m \geq 1$ arcs over chords of length $L_i = 1$ and perhaps one lower arc over a chord of length $0 < L_{m+1} < 1$. If any of the first m arcs were more than a semicircle, again the area under two would exceed $\pi/4$, a contradiction. Also a lower arc of the same curvature cannot exceed a semicircle. Hence for $1 < L_0$ we are in the convex range of arc. The length of each arc is $L_i \text{arc}(x_i/L_i^2)$. Since each $L_i \leq 1$, the total length

$$\sum L_i \text{arc}(x_i/L_i^2) \geq \sum L_i \text{arc}(x_i/L_i) \geq L_0 \text{arc}(X/L_0)$$

by convexity, as desired.

Finally we show that $L_0 \text{arc}(|X|/L_0)$ is increasing, or equivalently that $\text{arc}(x)/x$ is decreasing, or equivalently that $\text{arc}'(x) < \text{arc}(x)/x$. Since $\text{arc}(x)$ is convex for $0 \leq x \leq \pi/8$ and concave for $\pi/8 \leq x$ as in Figure 15.5.2, it suffices to check this at $\pi/8$ (the semicircle of radius 1/2), where indeed the curvature $\text{arc}'(\pi/8) = 2 < (\pi/2)/(\pi/8)$. ∎

15.6 Proposition (Hales Theorem 2) *Let C be a cluster of n regions of area at most 1 in R^2. Then the ratio of perimeter to area is greater than the ratio of perimeter to area for the unit regular hexagonal tiling.*

Proof More generally, without assuming that each area $A_i \leq 1$, we bound the ratio of perimeter to $\Sigma \min \{A_i, 1\}$. By considering connected components, we may assume that the regions are connected. To obtain a contradiction, take a counterexample to minimize the number of (connected) regions. We may assume $n \geq 2$ since the perimeter-to-area ratio for a circle of unit area is greater than the ratio for the hexagonal tiling, which is half the ratio for a regular hexagon.

By trivial modifications we may always assume that C is a smooth connected graph with no loops, all vertices of degree at least three, and simply connected faces. We may assume that C is simply connected by incorporating any bounded components of the exterior into the regions of C by removing edges. Each region with, for example, N edges must have area at least $2\pi/N^2\sqrt{3}$, or removing its longest edge would yield a counterexample with fewer regions (for details, see Hales Remark 2.7). If for the moment we think of C as lying on a sphere, then the number of vertices, edges, and regions satisfy

$$2 = v - e + n \qquad \text{(including the exterior)}.$$

Each region with N edges contributes at most $N/3$ to v and exactly $N/2$ to e because, for example, each edge bounds two regions (or possibly the same region twice). Therefore,

$$2 \leq \sum (N/3 - N/2 + 1) = \sum (1 - N/6) \qquad \text{(including the exterior)}.$$

If we remove the contribution of the exterior region, we have

$$1 \leq \sum (1 - N/6) \qquad \text{(without the exterior)}.$$

Hence, if we sum the hexagonal isoperimetric inequality 15.4(1) over all regions, the sum of the $-c(N - 6)$ terms is positive. The sum of the a_i terms from interior edges is 0 because the two regions sharing an edge make opposite contributions. The length P_i of each interior edge contributes twice to the sum of the perimeters. Therefore, summing (1) yields the proposition except for two discrepancies: the favorable undercounting of exterior perimeter P_i and the possibly unfavorable contribution of any exterior $.5a_i > 0$. But by Dido's inequality L_D,

$$P_i \geq \sqrt{2\pi a_i} = a_i \sqrt{2\pi/a_i} \geq a_i \sqrt{4\pi} > .5 a_i P_0.$$

because $a_i \leq 1/2$ and $\sqrt{4\pi} < .5 P_0 = .5 \sqrt[4]{12}$. ∎

Proposition 15.6 now has the following corollary.

15.7 The Hexagonal Honeycomb Theorem *Let C be a planar cluster of infinitely many regions of unit area. Then the limiting perimeter to area ratio*

$$\rho = \limsup_{r \to \infty} \frac{P(r)}{A(r)}$$

satisfies $\rho \geq \rho_0$, where ρ_0 is the ratio for the unit regular hexagonal tiling.

Proof We actually prove the result for regions of area at most 1. Then by considering the connected components, we may assume that the regions are connected. By the Truncation Lemma 15.3, ρ is greater than or equal to a limit of ratios for finite truncations. Therefore, by Proposition 15.6, $\rho \geq \rho_0$. ∎

15.8 The Bees' Honeycomb The bees actually have a more complicated, three-dimensional problem involving how the ends of the hexagonal cells are shaped to interlock with the ends of the cells on the other side. L. Fejes Tóth [5], in his famous article on "What the bees know and what they do not know," showed that the bees' three-dimensional structure can be improved slightly, at least for the mathematical model with infinitely thin walls. See Thompson [pp. 107–119] for interesting history and discussion on "The Bee's Cell."

15.9 Unequal Areas One can generalize the planar partitioning problem to finitely many prescribed areas $A_1, \ldots, A_k$ and prescribed probabilities $p_1 + \cdots + p_k = 1$ such that as $r \to \infty$ the fraction of regions within $\mathbf{B}(0, r)$ of area A_i approaches p_i. For approximately equal areas, the minimizer probably consists essentially of regular hexagons, as G. Fejes Tóth has shown can be accomplished by partitions of the plane. Figure 15.9.1 suggests two other candidate minimizers for

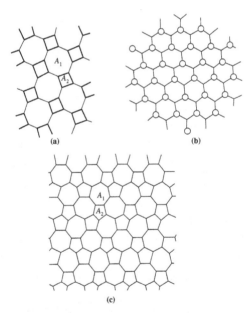

Figure 15.9.1. Candidate minimizers for two prescribed areas $A_1 > A_2$. The structures should relax to circular arcs meeting at 120-degree angles. (From Morgan [8, Figure 2], based on figure from Branko Grünbaum and G. C. Shephard, "Tilings and Patterns," Freeman and Co., New York, 1987).

two equally likely prescribed areas ($k = 2$, $A_1 > A_2$, $p_1 = p_2 = 1/2$), which do better than regular hexagons. Figure 15.9.1a does better approximately for $.117 < A_2/A_1 < .206$, and Figure 15.9.1c does better approximately for $.367 < A_2/A_1 < .421$. Another arrangement of 4-gons and 8-gons does better approximately for $.044 < A_2/A_1 < .099$. Sometimes separating or "sorting" the larger bubbles from the smaller ones beats mixing them together. See Teixeira, Graner, and Fortes [2] and references, including wider exploration for discs by Likos and Henley.

15.10   Kelvin Conjecture Disproved by Weaire and Phelan

1994 brought striking news of the disproof of Lord Kelvin's 100-year-old conjecture by Denis Weaire and Robert Phelan of Trinity College, Dublin. Kelvin sought the least-area way to partition all of space into regions of unit volumes. (Since the total area is infinite, least area is interpreted to mean that there is no area-reducing alteration of compact support preserving the unit volumes.) His basic building block was a truncated octahedron, with its 6 square faces of truncation and 8 remaining hexagonal faces, which packs perfectly to fill space as suggested by Figure 15.10.1. (The regular dodecahedron, with its 12 pentagonal faces, has less area, but it does not pack.) The whole structure relaxes slightly into a curvy equilibrium, which is Kelvin's candidate. All regions are congruent.

Kelvin loved this shape, constructed models, and exhibited stereoscopic images as in Figure 15.10.2.

Weaire and Phelan recruited a crystal structure from certain chemical "clathrate" compounds, which uses two different building blocks: an irregular dodecahedron

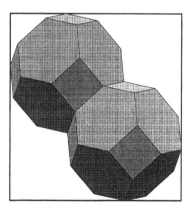

Figure 15.10.1. Lord Kelvin conjectured that the least-area way to partition space into unit volumes uses relaxed truncated octahedra. Graphics by Ken Brakke in his Surface Evolver from Brakke's early report [5].

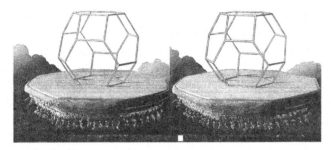

Figure 15.10.2. Kelvin loved his truncated octahedron, constructed models, and exhibited the pictured stereoscopic images. (Crossing your eyes to superimpose the two images produces a three-dimensional view.) [Thomson, p. 15.]

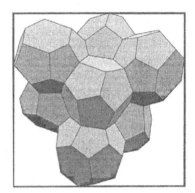

Figure 15.10.3. The relaxed stacked tetrakaidecahedra and occasional dodecahedra of Weaire and Phelan beat Kelvin's conjecture by approximately 0.3%. Graphics by Ken Brakke in his Surface Evolver from Brakke's early report [5].

and a tetrakaidecahedron with 12 pentagonal faces and 2 hexagonal faces. The tetrakaidecahedra are arranged in three orthogonal stacks, stacked along the hexagonal faces, as in Figure 15.10.3. The remaining holes are filled by dodecahedra. Again, the structure is allowed to relax into a stable equilibrium. Computation in the Brakke Evolver shows an improvement over Kelvin's conjecture of approximately 0.3%. The rigorous proof, by Kusner and Sullivan [1], proves only approximately 0.01%.

In greater detail, the centers of the polyhedra are at the points of a lattice with the following coordinates modulo 2:

0	0	0
1	1	1
0.5	0	1
1.5	0	1
0	1	0.5
0	1	1.5
1	0.5	0
1	1.5	0

Given a center, the corresponding polyhedral region is just the "Voronoi cell" of all points closer to the given center than to any other center. The relaxation process also needs to slightly adjust the volumes to make them all 1.

Weaire and Phelan thus provided a new conjectured minimizer. Weaire's popular account in *New Scientist* gives further pictures and details (see also Klarreich). Don't miss the pictures of Kelvin, Weaire, and Phelan in *Discover Magazine's* comic-book version by Larry Gonick.

15.11 Higher Dimensions Kelvin's truncated octahedron is actually a scaled "permutohedron," the convex hull of the 24 permutations of $(1, 2, 3, 4)$ in $\mathbf{R}^3 = \{\mathbf{x} \in \mathbf{R}^4 : \Sigma x_i = 10\}$. Likewise, the regular hexagon is the permutohedron in $\mathbf{R}^2 = \{\mathbf{x} \in \mathbf{R}^3 : \Sigma x_i = 6\}$. Will permutohedra turn out to relax into optimal partitions in higher dimensions? To the contrary, John M. Sullivan (personal communication, 1999) conjectures that the least-area partitioning of $\mathbf{R}^4$ into unit volumes is given by what Coxeter [Chapter VIII] calls the regular honeycomb $\{3, 4, 3, 3\}$, with octaplex cells $\{3, 4, 3\}$, each with 24 octahedral faces. Intriguingly, this polyhedral foam has no need of relaxation. Its surfaces and lines all meet at the ideal angles. Likewise in $\mathbf{R}^8$, there is such a candidate based on the so-called E8 lattice. See Conway and Sloane.

15.12 How the Weaire–Phelan Counterexample to the Kelvin Conjecture Could Have Been Found Earlier The clathrate compounds that inspired Weaire and Phelan had just 3 years earlier inspired counterexamples by Tibor Tarnai to related conjectures on the optimal way to cover a sphere with discs (see Stewart). As early as 1890, J. Dana described similar structures in volcanic lava, more recently observed in popcorn (see Cashman *et al.*).

K. Brakke spent hours at his father's old desk seeking counterexamples. Had he reached up and pulled down his father's copy of Linus Pauling's classic, *The Nature of the Chemical Bond*, it would doubtless have fallen open to the illustration, in the clathrate compound section, of the chlorine hydrate crystal, essentially the Weaire–Phelan counterexample.

R. Williams, after spending years seeking a Kelvin counterexample, finally gave up and later published a well-illustrated *The Geometrical Foundation of Natural Structure: A Source Book of Design*. In his Figure 5.22, he pictured the Weaire–Phelan counterexample without realizing it. Of course, it would have been difficult to check without Brakke's Surface Evolver.

In 1988 at the Geometry Center, John M. Sullivan, inspired by Fred Almgren, computed Voronoi cells of equal volumes, but Weaire–Phelan requires weighted Voronoi cells (with the distance to each point weighted differently).

15.13 Conjectures and Proofs Optimal partitioning, as described in this chapter, is much more difficult than optimal packing or covering. Hales's proof for the hexagonal honeycomb in $\mathbf{R}^2$ did not come until 1999. Will the new Weaire–Phelan candidate for $\mathbf{R}^3$ take another century to prove?

Exercise

15.1 The perimeter of a finite cluster equals half the sum of the perimeters of the regions and the perimeter of their union. Moreover, the perimeter of a region is at least the perimeter of a round disc of the same area. Use these two facts to give a short proof of Proposition 15.6 for $n \leq 398$.

Immiscible Fluids and Crystals

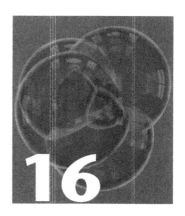

16.1 Immiscible Fluids Clusters of immiscible fluids $F_1, \ldots, F_m$ (with ambient F_0) such as oil, water, and mercury in air tend to minimize an energy proportional to surface area, where now the constant of proportionality $a_{ij} > 0$ depends on which fluids the surface separates. We might as well assume triangle inequalities $a_{ik} \leq a_{ij} + a_{jk}$, since otherwise an interface between F_i and F_k could profitably be replaced by a thin layer of F_j.

16.2 Existence of Minimizing Fluid Clusters The existence of least-energy clusters of immiscible fluids in $\mathbf{R}^n$ follows as for soap bubble clusters (13.4). Technically, it is very convenient to use flat chains with "fluid" rather than integer coefficients, after Fleming [1], White [2], and Morgan [9], so that, for example, the superposition of an oil–water interface and a water–mercury interface is automatically an oil–mercury interface.

16.3 Regularity of Minimizing Fluid Clusters Now suppose strict triangle inequalities hold. White [6; 2, Section 11] has announced that minimizing fluid clusters consist of smooth constant-mean-curvature hypersurfaces meeting along a singular set of Hausdorff dimension at most $n-2$. A form of monotonicity holds [Morgan 9, Section 3.2]. G. Leonardi has proved that if two fluids occupy most of a ball in a minimizer, then they occupy all of a smaller ball.

The structure of the singularities is not as well understood as for soap bubble clusters. In planar singularities, many circular arcs can meet at an isolated point. A conjecture on the classification of energy-minimizing cones in $\mathbf{R}^2$ was proved for up to five fluids and disproved for six by the Williams College NSF "SMALL" undergraduate research Geometry Group [Futer *et al.*].

Examples of energy-minimizing cones in $\mathbf{R}^3$ include cones over the edges of any tetrahedron, a regular polygonal prism, the regular octahedron, the regular dodecahedron, the regular icosahedron, and the cube; if, however, you move just

Geometric Measure Theory. http://dx.doi.org/10.1016/B978-0-12-804489-6.00016-2

one vertex of the cube, the cone is not minimizing for any choice of weights a_{ij} (Morgan, unpublished). It is an open question whether every combinatorial type of polyhedron occurs as an energy-minimizing cone.

16.4 Crystals (see Taylor [2], Feynman, Morgan [16, Chapter 10])

Clusters of crystals $C_1, \ldots, C_m$ tend to minimize an energy that depends on direction as well as the pair of crystals separated, an energy given by norms Φ_{ij} (cf. 12.5), sometimes assumed to be even so that $\Phi_{ij}(-v) = \Phi_{ij}(v)$.

For a single crystal of prescribed volume, the unique energy minimizer is the well-known Wulff shape (ball in dual norm; see Morgan [16, Chapter 10] and references therein). Typically, the norm is not smooth, certain directions are much cheaper than others, and the Wulff shape is a polytope, like the crystals of Figure 16.4.1. For salt, horizontal and vertical axis–plane directions are cheap and the Wulff shape is a cube.

Figure 16.4.1. Crystal shapes typically have finitely many flat facets corresponding to surface orientations of low energy. (The first two photographs are from Steve Smale's *Beautiful Crystals Calendar.* The third photograph is from E. Brieskorn. All three appear in *The Parsimonious Universe* by S. Hildebrandt and A. Tromba [pp. 263–264] and in Morgan [16, Figure 10.1]).

If the energy functional is lower semicontinuous, then the existence of least-energy clusters of crystals in $\mathbf{R}^n$ follows as for soap bubble clusters (13.4). Ambrosio and Braides [Example 2.6] (see Morgan [10]) show that the triangle inequalities $\Phi_{ik} \leq \Phi_{ij} + \Phi_{jk}$ do not suffice to imply lower semicontinuity. There are no general regularity or monotonicity results. For the case when the Φ_{ij} are multiples $a_{ij}\Phi$ of a fixed norm, Leonardi [Remark 4.7] has proved that almost every point in the surface has a neighborhood in which the surface is a minimizing interface between two regions (for fixed volumes).

16.5 Planar Double Crystals of Salt For the case of double crystals of salt in $\mathbf{R}^2$, Morgan, French, and Greenleaf proved that three types of minimizers occur, as in Figure 16.5.1. (Here, the energy norm is given by $\Phi(x, y) = |x| + |y|$ so that horizontal and vertical directions are cheap.) Electron microscope photographs of table salt crystals in Figure 16.5.2 show similar shapes. Undergraduates Wecht, Barber, and Tice considered a refined model in which the interface cost is some fraction λ of the external boundary cost. They found the same three types (with different dimensions), except that when $\lambda \leq \lambda_0 \approx .56$, the minimizer jumps directly from type I to type III (as shown in Figure 16.5.3).

16.6 Willmore and Knot Energies Physical rods or membranes may tend to minimize nonlinear elastic energies such as the Willmore energy $\int H^2$ (where H is the mean curvature). In 1965 [Willmore, Section 7.2] showed that the round sphere minimizes this energy. The Willmore Conjecture says that among tori, a standard "Clifford" torus of revolution (with circular cross section) minimizes energy. L. Simon [2] proved existence and regularity for a minimizing torus.

Kusner and Sullivan [2] studied a modified electrostatic energy on knots and links and provided stereoscopic pictures of (nonrigorously) computed minimizers.

16.7 Flows and Crystal Growth Energy minimization guides not only equilibrium shapes but also dynamical processes such as crystal growth. Length or area minimization gives rise to flows in the direction and magnitude of the curvature. Matt Grayson proved that a smooth Jordan curve in a closed surface flows by

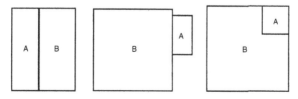

Figure 16.5.1. The three types of minimizing double crystals of salt.

Figure 16.5.2. Electronic microscope photographs of table salt crystals show shapes similar to those of Figure 16.5.1. [Morgan, French, and Greenleaf, Figure 2.]

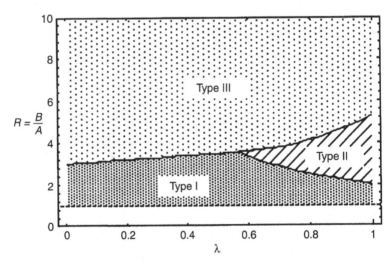

Figure 16.5.3. If the interface carries just a fraction λ of the exterior boundary cost, the transition to type III occurs sooner, without passing through type II when $\lambda \leq \lambda_0 \approx .56$. [Wecht, Barber, and Tice, Figure 4.]

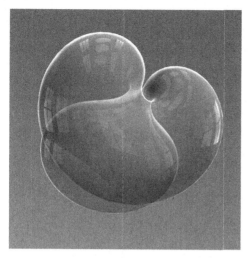

Figure 16.7.1. The Optiverse, a video by Sullivan, Francis, and Levy, begins its optimal way to turn a sphere inside out. http://new.math.uiuc.edu/optiverse.

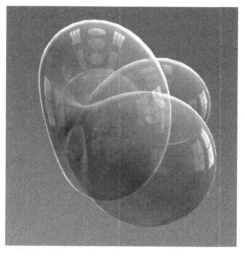

Figure 16.7.2. The saddle configuration halfway through the Optiverse sphere eversion. http://new.math.uiuc.edu/optiverse.

curvature to a round point or a geodesic. Gerhard Huisken proved that a smooth uniformly convex hypersurface in $\mathbf{R}^n$ flows by mean curvature to a round point. (Some nonconvex surfaces, such as a dumbbell, can develop singularities.) Other factors such as heat flow further complicate crystal growth. There are many definitions of flow, which agree in the nicest cases (see the survey on "Geometric models of crystal growth" by Taylor, Cahn, and Handwerker). Some recent approaches require the full strength of geometric measure theory (see Almgren, Taylor, and Wang). Hubert Bray used a simultaneous flow of surfaces and metrics in his proof of the Riemannian Penrose Conjecture in general relativity, which states roughly that the square of the mass of the universe is at least as great as the sum of the squares of the masses of its black holes. This result uses and generalizes the positive mass theorem of Schoen and Yau, which simply states that the mass of the universe is positive.

Francis, Sullivan, Kusner, *et al.*, used elastic energy gradient flow on Brakke's Surface Evolver to turn a sphere inside out,. starting as in Figure 16.7.1. The flow is downward in two directions from the halfway energy saddle point shown in Figure 16.7.2.

Perelman used the "Ricci flow" to prove the Poincaré Conjecture (see Chapter 18).

16.8 The Brakke Evolver Ken Brakke has developed and maintained magnificent software for the computer evolution of surfaces, from soap films to crystal growth, with beautiful graphics. It was Brakke's Evolver that established the Weaire–Phelan counterexample to the Kelvin Conjecture (Chapter 15). It was used to help redesign the fuel tanks of the Space Shuttle.

A package containing the source code, the manual, and sample data files is freely available at the Evolver web site at www.susqu.edu/facstaff/b/brakke/evolver/.

Isoperimetric Theorems in General Codimension

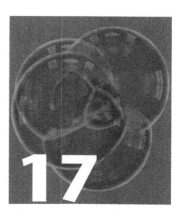

The classical isoperimetric inequality (Section 13.2) for the volume of a region in $\mathbf{R}^n$ in terms of its perimeter, maximized by the round ball, has important generalizations to higher codimension and to other ambients. It was not until 1986 that the classical isoperimetric inequality was extended to general codimension by Fred Almgren (Figure 17.0.1). Whereas in codimension 0 there is a unique region with given boundary, in higher codimension there are many surfaces (of unbounded area) with given boundary, and the isoperimetric inequality applies only to the one of least area.

17.1 Theorem (Almgren [2]) *An m-dimensional area-minimizing integral current in $\mathbf{R}^n$ ($2 \leq m \leq n$) has no more area than a round disc of the same boundary area, with equality only for the round disc.*

Remark Previously for $m < n$ this result was known just for $m = 2$. In that case the boundary, a curve in $\mathbf{R}^n$, may by decomposition be assumed to be a closed curve through the origin. The cone over the curve can be developed in the plane and thus shown to have no more area than a round planar disc (see Federer 4.5.14).

Proof Sketch As in 13.7, among area-minimizing surfaces with the same area as the round unit disc, there exists a Q of least boundary area.

The second step is to show that ∂Q has mean curvature at most 1 (the mean curvature of the boundary of the unit disc). Since ∂Q is not known to be smooth, this statement is formulated weakly, in terms of the first variation of ∂Q (see Section 11.2).

To make the ensuing discussion specific, we treat the case of a two-dimensional area-minimizing surface Q of area π in $\mathbf{R}^3$, chosen to minimize the length of its boundary curve C. Also, we suppress some relatively routine concerns about singularities.

Geometric Measure Theory. http://dx.doi.org/10.1016/B978-0-12-804489-6.00017-4

Figure 17.0.1. (a) Fred Almgren always presented his favorite ideas with generous enthusiasm. (b) Fred joyfully waves his honorary flag at a celebratory dinner. Seated on the near side of the table, starting on the right, are Aaron Yip, Jenny Kelley, David Caraballo, Karen Almgren, Jean Taylor, Elliott Lieb, and Gary Lawlor (with Ed Nelson barely visible between and behind Lawlor and Lieb). At the head of the table and then along the far side are Frank Morgan, Joe Fu, Mohamed Messaoudene, Melinda Duncan, Andy Roosen, Fred Almgren, Christiana Lieb, and Dana Mackenzie. Photos by Harold Parks, courtesy of Parks. These photos also appeared in the Almgren memorial issue of the *Journal of Geometric Analysis* 8 (1998).

Note that flowing the boundary of the unit disc (with unit curvature) by its curvature and then rescaling to the original enclosed area just restores the unit disc, with no change in enclosed area. Moreover, this flow, directly into the surface, realizes a general upper bound on the reduction of least area enclosed; any smaller

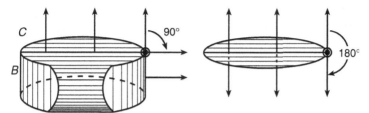

Figure 17.1.1. The boundary B of the convex hull of the curve C is generally flat in one direction so that the entire area 4π of the image of the Gauss map is due to the singular contribution along $B \cap C$, depending on a dihedral angle α, here often 90 degrees, whereas it is 180 degrees for a flat circle.

new minimizer, together with the strip of surface swept out by the flowing boundary, would have less area than the original minimizer, a contradiction. By comparison, if C somewhere had curvature greater than 1, a local variation followed by rescaling would reduce the length of C, a contradiction.

The third step is to show that the length of a curve with curvature bounded by 1 is at least 2π, with equality only for a round circle. Consider the Gauss map G mapping each point of the boundary B of the convex hull of C to the unit sphere. Because $B - C$ is generally flat in one direction as in Figure 17.1.1, the entire area 4π of the unit sphere is due to the singular contribution along $B \cap C$, where the surface has a dihedral angle $0 \le \alpha \le \pi$. The contribution of an element ds to the area 4π of the sphere is proportional to $\kappa \alpha \, ds$. Consideration of the example of the unit circle with $\kappa = 1$ and $\alpha = \pi$ shows the constant to be $2/\pi$. Therefore,

$$4\pi = \frac{2}{\pi} \int_C \kappa \alpha \, ds \le \frac{2}{\pi} 1\pi |C|$$

so that $|C| \ge 2\pi$, as desired.

If equality holds, then $\alpha = \pi$ so that the convex hull has no interior and must be planar; moreover, $\kappa = 1$, and C must be a round unit circle.

The following generalization of the classical Minkowski inequality by Almgren and Allard gives a general isoperimetric inequality for m-dimensional minimal (stationary, not necessarily area-minimizing) surfaces (possibly with singularities) in $\mathbf{R}^n$. It is conjectured that the round disc is the extreme case, but this remains open even for smooth minimal surfaces of several boundary components in $\mathbf{R}^3$; see Section 4 of the survey on "The isoperimetric inequality" by Osserman [1], and [Sullivan and Morgan, Problem 22].

17.2 Theorem (Allard Theorem 7.1 and Corollary 7.2) *There is a constant $c(m, n)$ such that an m-dimensional minimal submanifold with boundary of $\mathbf{R}^n$ has no more than c times the area of a round disc of the same boundary area. More generally, any submanifold M with boundary and mean curvature vector $\mathbf{H}$ satisfies*

$$(1) \qquad |M|^{(m-1)/m} \leq c \left(|\partial M| + \int_M |\mathbf{H}| \right).$$

Still more generally, any compact integral varifold V satisfies

$$(2) \qquad |V|^{(m-1)/m} \leq c|\delta V|.$$

Here $|\ \ |$ is used for total area or measure.

Remark The proof uses monotonicity (and a covering argument), and hence does not generalize to more general integrands. Of course, *minimizers* for general integrands satisfy certain inequalities by comparison with minimizers for area.

17.3 General Ambient Manifolds As described in Section 12.3, Federer provides a very general isoperimetric inequality for area-minimizing surfaces in a general Riemannian manifold M with boundary, with little information about how the constant depends on M. Hoffman and Spruck [Theorem 22] give a delicate generalization of 17.2(1) to small submanifolds of a Riemannian manifold, depending on the mean curvature of the submanifold and the sectional curvature of the ambient.

In codimension 0, Yau [Section 4] (using divergence of the distance function) and Croke (using integral geometry) give linear and standard isoperimetric inequalities depending on diameter, volume, and Ricci curvature.

The *Aubin* or *Cartan-Hadamard Conjecture* seeks sharp isoperimetric inequalities in a simply connected Riemannian manifold of sectional curvature $K \leq K_0 \leq 0$. This has been proved only in dimension two, and in dimension three by Bruce Kleiner by a generalization of Almgren's proof of Theorem 17.1; see Morgan and Johnson.

Manifolds with Density and Perelman's Proof of the Poincaré Conjecture

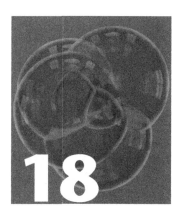

Perelman's 2003 proof of the 1904 Poincaré Conjecture considers a manifold with a density (as in freshman physics or calculus), as we will explain. Manifolds with density long have appeared in mathematics, with more recent attention to their differential geometry, including a generalization of Ricci curvature, which Perelman uses in exploring the Ricci flow. There are useful isoperimetric results, such as a generalization of the Levy–Gromov isoperimetric comparison theorem (18.7). In the 1980s, Bakry and Ledoux used Markov semigroup arguments. Bayle (18.9) used first and second variation formulas. The approach here, following and correcting our survey [Morgan 25, 26] via Heintze–Karcher, seems to be the simplest. The goal is to generalize all of Riemannian geometry to manifolds with density. For the explosion of results since 2003, including singular manifolds with density (metric measure spaces), see Morgan [29].

18.1 Definitions A *manifold with density* is a Riemannian manifold M^n with a positive density function $\Psi(x)$ used to weight volume and hyperarea (and sometimes lower-dimensional area and length). In terms of the underlying Riemannian volume dV_0 and area dA_0, the new, weighted volume and area are given by

$$dV = \Psi \, dV_0, \quad dA = \Psi \, dA_0.$$

Such a density is not equivalent to scaling the metric conformally by a factor $\lambda(x)$ since in that case volume and area would scale by different powers of λ. Manifolds with density, a special case of the "mm spaces" of Gromov [2] or the earlier "spaces of homogeneous type" (see [Coifman and Weiss, pp. 587, 591]), long have arisen on an *ad hoc* basis in mathematics. Quotients of Riemannian manifolds are manifolds with density. For example, $\mathbf{R}^3$ modulo rotation about the z-axis is the half-plane

Geometric Measure Theory. http://dx.doi.org/10.1016/B978-0-12-804489-6.00018-6

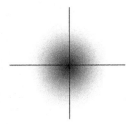

Figure 18.1.1. Gauss space is Euclidean space with Gaussian density, which drops off exponentially. Image by Diana Davis.

$$H = \{(x, z) : x \geq 0\}$$

with density $2\pi x$; volume and area in $\mathbf{R}^3$ are given by integrating this density over the generating region or curves in H. A manifold with density of much interest to probabilists is *Gauss space* G^n (see Ledoux–Talagrand, McKean, or Stroock), which is Euclidean space with Gaussian density

(1) $$\Phi = (2\pi)^{-n/2} e^{-x^2/2},$$

as in Figure 18.1.1. The leading coefficient just gives G^n unit weighted volume. Note that Gauss space is a product space and is rotationally invariant. Any such density on $\mathbf{R}^n$ must be a multiple of e^{cx^2}. The normalized scalings of Gauss space have density $(\gamma/2\pi)^{n/2} e^{-\gamma x^2/2}$.

In Gauss space, the isoperimetric problem has an even nicer solution than the hyperspheres of Euclidean space or spheres. Because of the rapidly decreasing density, a perimeter-minimizing surface could well be unbounded, and it is.

18.2 Theorem (Sudakov–Tsirel'son, Borell [1]) *In Gauss space, hyperplanes minimize area for given volume.*

Proof Sketch Mehler [Stroock, Exercise 2.1.40 p. 76, and footnote p. 77] observed that for fixed n, Gauss space G^n can be obtained as the limit of orthogonal projections into $\mathbf{R}^n$ of high-dimensional spheres $\mathbf{S}^{N-1}$ of radius $\sqrt{N}$, with density normalized to yield unit volume. To see this, first make the trivial observation that since G^N is a product space, G^n is the orthogonal projection of G^N for $N \geq n$. The unrelated components of a vector in G^N have Gaussian distributions (in other words, they are independent Gaussian random variables). An easy computation shows that the square of each component has expected value 1. By the central limit theorem, the sum of the squares is probably close to N; that is, the measure is (uniformly) concentrated on the sphere of radius $\sqrt{N}$. Thus, G^n is the limit of projections of such spheres.

The inverse projection of a hyperplane in G^n is a hypersphere in $\mathbf{S}^{N-1}$. Since hyperspheres are known to be area minimizing for given volume in $\mathbf{S}^{N-1}$, a careful limit argument shows that hyperplanes are area minimizing in G^n.

Theorem 18.7 and Corollary 18.10 provide two other proofs that also show uniqueness, first proved by Carlen and Kerce. Actually the simplest proof uses a version of Steiner symmetrization for Gauss space called Ehrhard symmetrization, in which slices by lines are replaced by half-lines instead of by centered intervals (see Cianchi *et al.*).

The fact that not only Gauss space but also its isoperimetric regions are products makes it an excellent model with implications for other spaces. For example, it happens to follow quickly from Theorem 18.2 that the least-perimeter way to divide a cube in half is a wall down the middle, a result first proved by Hadwiger by a clever, *ad hoc* argument. See Ros [1, §1.5] for this and other consequences and references.

For applications to Brownian motion and to stock option pricing, see Borell [1, 2].

18.3 Curvature   There is a canonical extension to smooth manifolds with (smooth) density e^ψ of the notion of the mean curvature of a smooth hypersurface: as in the classical case, it is the quantity that appears in the first variation formula (18.9) for how the area changes as the surface moves. The formula for the generalized mean curvature H_ψ involves both the classical mean curvature H and the log of the density:

$$(1) \qquad H_\psi = H - \frac{1}{n-1}\frac{d\psi}{dn}.$$

General regularity for minimizers extends to manifolds with density (§8.5). In particular, for a perimeter-minimizing surface for fixed volume, H_ψ must be constant since otherwise, by the first variation formula, moving the surface outward where H_ψ is larger and inward where H_ψ is smaller could preserve volume and decrease perimeter. As in classical differential geometry, $-(n-1)H_\psi$ has the interpretation dP/dV.

For a two-dimensional smooth manifold with density e^ψ, Corwin *et al.* define a generalized Gauss curvature

$$(2) \qquad G_\psi = G - \Delta\psi$$

and obtain a generalization of the Gauss–Bonnet formula for a smooth disc R:

$$(3) \qquad \int_R G_\psi + \int_{\partial R} \kappa_\psi = 2\pi,$$

where κ_ψ is the inward one-dimensional generalized mean curvature as in (1) and the integrals are with respect to unweighted Riemannian area and arc length.

Certain punctured planes with density with $G_\psi = 0$ were studied by Carroll *et al.* Different generalizations of Gauss curvature, involving $|\nabla \psi|^2$, are needed to recover asymptotic formulas for areas and perimeters of small discs [Corwin *et al.*, Props. 5.8 and 5.9]. On generalizations of sectional curvature, see e.g. Wylie.

In general dimensions, there are various useful generalizations of Ricci curvature (see Bayle, Villani, and references therein to Bakry, Émery, Ledoux, and others; also Chang *et al.*), generally involving Hess ψ and $d\psi \otimes d\psi$. My favorite generalization of Ricci curvature is simply

$$(4) \qquad \mathrm{Ric}_\psi = \mathrm{Ric} - \mathrm{Hess}\ \psi,$$

the generalized curvature of Lichnerowicz, Bakry–Émery [Prop. 3], and Bakry–Ledoux [p. 265]. It yields the simplest generalization of Levy–Gromov 18.7, appears in the second variation formula 18.9, and appears in Perelman's paper on the Poincaré Conjecture (18.11). For Gauss space G^n (1) with density $\Phi = e^\varphi$ it is constant:

$$\mathrm{Ric}_\varphi = 0 - \mathrm{Hess}\,\varphi = I.$$

Unlike the classical case, for a two-dimensional manifold with density, the Gauss curvature (2) is not quite half the trace of the Ricci curvature (4).

The classical Heintze–Karcher formula (see also [Burago–Zalgaller, 34.1.10(11)]) provides an upper bound on the volume of a one-sided neighborhood of a hypersurface in terms of its mean curvature and the Ricci curvature of the ambient manifold.

18.4   Theorem (Classical Heintze–Karcher)   *Let M^n be a smooth, complete Riemannian manifold with Ricci curvature $\mathrm{Ric} \geq (n-1)\delta$. Let S be a smooth, oriented, closed hypersurface in M with mean curvature $H(s)$. Let $V(r)$ denote the volume of the region within distance r of S on the side of the unit normal (which determines the sign of H). Then*

$$(1) \qquad V(r) \leq \int_S \int_0^{r^*(s)} [c_\delta(t) - H(s)s_\delta(t)]^{n-1} dt\, ds,$$

where ds denotes weighted surface area,

$$s_\delta(t) = \begin{cases} \delta^{-1/2} \sin \delta^{1/2} t & \text{for } \delta > 0, \\ t & \text{for } \delta = 0, \\ |\delta|^{-1/2} \sinh |\delta|^{1/2} t & \text{for } \delta < 0. \end{cases}$$

$c_\delta(t) = ds_\delta(t)/dt$, and r^* is the lesser of r and the first zero of $c_\delta(t) - H(s)s_\delta(t)$. The result generalizes to closed surfaces of higher codimension.

The following generalization of Heinze–Karcher to manifolds with density is simpler and has an easier proof.

18.5   Theorem (Heintze–Karcher for Manifolds with Density)

Let M^n be a smooth, complete Riemannian manifold with smooth density $\Psi = e^\psi$ satisfying

$$\mathrm{Ric}_\psi = \mathrm{Ric} - \mathrm{Hess}\ \psi \geq \gamma.$$

Let S be a smooth, oriented, complete, finite-area hypersurface in M with generalized mean curvature

$$H_\psi(s) = H(s) - \frac{1}{n-1}\frac{d\psi}{dn}.$$

Let $V(r)$ denote the volume of the region within distance r of S on the side of the unit normal (which determines the sign of H_ψ). Then

(1) $$V(r) \leq \int_S \int_0^r \exp(-(n-1)H_\psi(s)t - \gamma t^2/2)dtds,$$

where ds denotes weighted surface area.

If equality holds, then S has vanishing classical second fundamental form, the region is a metric product $S \times [0,r]$, and inside the region, along geodesics normal to S, $-d^2\psi/dt^2 = \gamma$.

Theorem 18.5 is sharp for hyperplanes in Gauss space, for example.

Proof   We begin with the case $\Psi = 1$. This case follows from Theorem 18.4, but we want to incorporate variable Ricci curvature, as in (3) below, and give an easier proof. Consider the volume element $e^{f(s,t)}dtds$, corresponding to an infinitesimal slice dt of an infinitesimal normal wedge from S. Then $e^{f(s,t)}ds$ represents an element of surface area parallel to S. By the first variation formula 18.9, for example, its derivative $f'e^f ds$ equals $-(n-1)He^f ds$, so that

$$f' = -(n-1)H.$$

Since $(n-1)H' = \mathrm{II}^2 + \mathrm{Ric}(n,n)$, where II is the second fundamental form (see the remark after Section 18.9),

$$f'' = -\mathrm{II}^2 - \mathrm{Ric}(n,n) \leq -\mathrm{Ric}(n,n),$$

with equality only if II^2 vanishes. Hence, by Taylor's theorem,

(2) $$f(s,t) \leq -(n-1)H(s)t - \int_0^t (t-\tau)\mathrm{Ric}(n,n)d\tau.$$

Consequently, since every point of $V(r)$ is covered by the infinitesimal wedge from the nearest point of S,

(3) $$V(r) \leq \int_S \int_0^r \exp\left(-(n-1)H(s)t - \int_0^t (t-\tau)\mathrm{Ric}(n,n)d\tau\right)dtds_0,$$

where we now write ds_0 to emphasize that this is the case of unweighted area.

For general density $\Psi = e^\psi$,

$$(4) \qquad \psi(s,t) \leq \psi(s,0) + t\frac{d\psi}{dn}(s,0) + \int_0^t (t-\tau)\frac{d^2\psi}{dn^2}d\tau.$$

Preparing to add f and ψ, note that

$$-(n-1)H(s) + t\frac{d\psi}{dn}(s) = -(n-1)H_\psi(s)$$

and that by hypothesis

$$-\int_0^t (t-\tau)\text{Ric}(n,n)d\tau + \int_0^t (t-\tau)\frac{d^2\psi}{dn^2}d\tau \leq -\gamma \int_0^t (t-\tau)d\tau = -\gamma t^2/2.$$

Hence,

$$f(s,t) + \psi(s,t) \leq \psi(s,0) - (n-1)H_\psi(s)t - \gamma t^2/2.$$

Therefore,

$$V(r) \leq \int_S \int_0^r e^f e^\psi dt\, ds_0$$

$$\leq \int_S \int_0^r \exp(-(n-1)H_\psi(s)t - \gamma t^2/2)dt(e^{\psi(s,0)}ds_0),$$

as desired. If equality holds, the parallel hypersurfaces are disjoint and have $II = 0$ so that the normal geodesics stay equidistant, which means that the region is a metric product. ∎

Remark Theorems 18.4 and 18.5 and their proofs apply to perimeter minimizers S with singularities. Indeed, for any point off S, the nearest point on S is a regular point because the tangent cone lies in a half-space and hence must be a hyperplane.

The classical isoperimetric theorem of Levy and Gromov ([Gromov 1, 2.2], [Burago–Zalgaller, 34.3.2], or [Ros 1, Sect. 2.5]) compares a compact manifold of positive Ricci curvature to the model sphere. The statement is simplified by using non-unit constant density to normalize to unit volume.

18.6 Theorem (Classical Levy–Gromov) *Let M^n be a smooth, complete, connected Riemannian manifold with constant density, unit volume, and Ricci curvature $\text{Ric} \geq \gamma > 0$. Then the isoperimetric profile $P(V)$ (least perimeter to enclose given volume) satisfies*

$$P \geq P_{S\gamma},$$

where $P_{S\gamma}$ is the isoperimetric profile of the (constant-density, unit-volume) round sphere of Ricci curvature γ.

Theorem 18.7 gives a generalization to a manifold M with variable density, with the sphere replaced by Gauss space as the model of comparison. (Gauss space beats the sphere.) It includes the sharp isoperimetric inequality for Gauss space, Proposition 18.2. If M has finite volume, we may assume by scaling the density that M has unit volume. Such scaling does not affect the generalized Ricci curvature because multiplying the density e^{ψ} by a constant just adds a constant to ψ and leaves Hess ψ unchanged.

18.7 Theorem (Levy–Gromov for Manifolds with Density) *Let M^n be a smooth, complete, connected Riemannian manifold with smooth density $\Psi = e^{\psi}$, unit volume, and generalized Ricci curvature*

(1) $$\mathrm{Ric}_{\psi} = \mathrm{Ric} - \mathrm{Hess}\ \psi \geq \gamma > 0.$$

Then the isoperimetric profile $P(V)$ (least perimeter to enclose given volume) satisfies

(2) $$P \geq P_{G\gamma},$$

where $P_{G\gamma}$ is the isoperimetric profile of (scaled) Gauss space with density

$$\Phi = e^{\varphi} = (\gamma/2\pi)^{n/2} e^{-\gamma x^2/2}$$

so that $\mathrm{Ric}_{\varphi} = -\mathrm{Hess}\ \varphi = \gamma$.

In Gauss space, perimeter minimizers are hyperplanes. If equality holds in (2) for some $0 < V < 1$, then M is a product of 1D Gauss space with some manifold with density.

Remark $P_{G\gamma}$ is independent of dimension, because the volume of a hyperplane bounding given volume is independent of dimension, because Gauss space is a product of 1D Gauss spaces. Hence, $P_{G\gamma}$ is just the value of the Gaussian density

$$\sqrt{\frac{\gamma}{2\pi}}\, e^{-\gamma x^2/2}$$

at the endpoint of a half-line with the given mass.

Gromov [2, 9.2] gives an alternative proof of the isoperimetric inequality in Gauss space as well as the sphere by integrating isoperimetric inequalities on slivers.

Proof Note that the generalized curvature H_{φ} of the hyperplane $S_0 = \{x_1 = a\}$ in Gauss space is given by

$$H_\varphi = H - \frac{1}{n-1}\frac{d\varphi}{dn} = 0 - \frac{1}{n-1}\gamma a,$$

which is constant (as it must be for any minimizer). For given $0 < V < 1$, a least-perimeter region of volume V exists by the compactness theorem, 5.1, 9.1 (since Gauss space has finite volume, there can be no volume loss to infinity in the limit). Let P be the perimeter given by such a minimizing hypersurface S in M and let P_0 be the perimeter given by the hyperplane S_0 in Gauss space. By replacing V by $1 - V$ (which changes the sign of the mean curvatures) if necessary, we may assume that the generalized mean curvature of S is greater than or equal to that of S_0. By generalized Heintze–Karcher (Theorem 18.4),

$$\frac{V}{P} \leq \frac{V}{P_0} = \int_0^\infty \exp(\gamma at - \gamma t^2/2)dt.$$

Taking M to be Gauss space, we conclude that hyperplanes are perimeter minimizing, then equality holds in 18.5(1) for $r = \infty$ on both sides of S, S has vanishing classical second fundamental form in the underlying Euclidean space, and hence S is a hyperplane. We conclude that in Gauss space, hyperplanes are uniquely perimeter minimizing.

Returning to general M, we conclude that $P \geq P_{G\gamma}$. If equality holds, the parallel hypersurfaces are disjoint and totally geodesic ($\mathrm{II}^2 = 0$) so that the normal geodesics stay equidistant, which means that the region is a metric product. ∎

The theorem of (Bonnet and) Myers (or see Chavel [2, Thm. 2.12] or Morgan [16, Thm. 9.6]) states that the diameter of a smooth, connected, complete, nD Riemannian manifold with Ricci curvature at least $a > 0$ is at most $\pi\sqrt{(n-1)/a}$. My favorite generalization to manifolds with density uses a different replacement for the Ricci curvature:

18.8 Myers' Theorem with Density [Morgan 26] *Let M^n be a smooth, connected Riemannian manifold with smooth density $e^\psi \leq b$, complete in the weighted metric. If (in the unweighted metric)*

$$\mathrm{Ric} - \Delta\psi + \mathrm{Hess}\ \psi \geq a > 0,$$

then the weighted diameter satisfies

$$\mathrm{diam} \leq \pi b\sqrt{(n-1)/a}.$$

The proof is a minor modification of the standard proof of Myers' theorem. Since the only weighted quantity to appear in the statement is length, there is (another, different) such result by applying the classical Myers' theorem to the conformally altered metric, but the hypothesis in terms of the original metric becomes unwieldy. Qian [Thm. 5] (cf. Bayle [E2.1, p. 233]) provides other generalizations which depend on $|\nabla \psi|^2$. See also Villani [Proposition 29.11], Wei-Wylie [Thms. 5.2, 1.3], and references therein. There are also various volume estimates and generalizations of Bishop's theorem.

The alternative approach of Bayle, similar in essence to Bakry–Émery, uses just second variation. For the record, we present such formulas for manifolds with density. The classical formulas are augmented by terms involving the first and second normal derivatives of ψ. For a second variation formula for fixed volume, see Proposition 20.1.

18.9 First and Second Variation [Bayle, Sect. 3.4.6] *Let M^n be a smooth Riemannian manifold with smooth density $\Psi = e^\psi$. Let S be a smooth hypersurface and consider a smooth normal variation of compact support of constant velocity $u(s)$ along the geodesic normal to S at s. Then the first and second variations of (weighted) area satisfy*

(1) $$\delta^1(u) = -\int_S u(n-1)H_\psi,$$

(2) $\delta^2(u) = \int_S |\nabla u|^2 + u^2(n-1)^2 H_\psi^2 - u^2 II^2 - u^2 \mathrm{Ric}(n,n) + u^2(d^2\psi/dn^2),$

where II is the second fundamental form and the generalized mean curvature H_ψ satisfies

$$H_\psi = H - \frac{1}{n-1}\frac{d\psi}{dn}.$$

The integrals are with respect to weighted area.

Remark If $u = 1$, then the second variational formula is equivalent to the first plus the fact that

$$(n-1)\frac{dH}{dt} = II^2 + \mathrm{Ric}(n,n),$$

which follows from the case of the curvature κ of a curve in a surface of Gauss curvature G:

$$\frac{d\kappa}{dt} = \kappa^2 + G.$$

18.10 Corollary [Bayle, (3.40)] *Let M^n be a smooth, complete, finite-volume Riemannian manifold with smooth density $\Psi = e^\psi$ with*

$$\mathrm{Ric} - \mathrm{Hess}\ \psi \geq \gamma.$$

Then the isoperimetric profile P and its derivatives P', P'' satisfy

(1) $$P P'' \leq -\gamma$$

almost everywhere (and in a weak sense everywhere). If equality holds, then a perimeter minimizer has vanishing classical second fundamental form.

In Gauss space, equality holds and perimeter minimizers are hyperplanes.

By scaling the density, which does not affect Hess ψ or PP'' or hence the hypothesis or conclusion of Corollary 18.10, one may assume that M has unit volume. Since equality holds in (1) for Gauss space, isoperimetric estimates such as Levy–Gromov (Theorem 18.7) follow; here, for example, we deduce the isoperimetric inequality in Gauss space.

Proof When we say that (1) holds in a weak sense, we mean that at each point, $P(V)$ lies under a smooth function with the same value and second derivative at most $-\gamma/P$. To prove this in the case that a perimeter-minimizing hypersurface is smooth, as always holds if $n \leq 7$ (§8.5), we just consider uniform normal perturbations, whose perimeters $P(V)$ give an upper bound on the least perimeter. To compute $P'(V)$ and $P''(V)$ for this new family, we use the variation formulae with $u = 1$. Using dots for time derivatives, by the first variation formula, since $-(n-1)H_\psi = P'$,

$$\dot{P} = PP'.$$

In the second variation formula, the first term of the integrand vanishes. The second term is P'^2. The third term is nonpositive and vanishes if and only if $\text{II} = 0$. The last two terms are by hypothesis at most $-\gamma$. Therefore,

$$\ddot{P} \leq P(P'^2 - \gamma).$$

Since $\dot{V} = P$, by the chain rule $P' = dP/dV = \dot{P}/P$ and

$$P'' = \frac{\ddot{P}P - \dot{P}^2}{P^3} \leq \frac{P^2(P'^2 - \gamma) - P^2 P'^2}{P^3} = \frac{-\gamma}{P}.$$

Therefore, $PP'' \leq -\gamma$, and equality implies that the perimeter minimizer has $\text{II} = 0$, as desired.

Even if the perimeter-minimizing surface is not smooth, the singular set is very small, of codimension at least 8 (§8.5), and one can use limits of variations u that vanish in a small neighborhood of the singular set; for details see [Morgan and Ritoré, Lemma 3.1].

To prove minimizers in Gauss space, let $P_0(V)$ denote the perimeter of a hyperplane enclosing volume V, for which equality holds in (1) with $\gamma = 1$ (Exercise 18.8). Suppose that for some V_0, $P(V_0) < P_0(V_0)$ and that V_0 is chosen to maximize $P_0(V_0) - P(V_0)$. Since $P''(V_0) \leq -1/P(V_0) < -1/P_0(V_0) = P_0''(V_0)$, this yields an immediate contradiction. Hence, $P = P_0$, equality always holds in (1), and hyperplanes are minimizers. Moreover, any minimizer has $\mathrm{II} = 0$. Since the underlying Riemannian manifold is Euclidean, any minimizer is a hyperplane.   ∎

Remarks For unit or constant density, 18.10(1) is not sharp for the model case of the sphere. In that category, there is a sharp result [Morgan and Johnson, Prop. 3.3]:

(2) $$P P'' \leq -P'^2/(n-1) - \gamma,$$

obtained by keeping the II^2 term and using the estimate $\mathrm{II}^2 \geq (n-1)H^2$, which does not hold in the category of manifolds with density for H_ψ.

Unfortunately, formulas (1) and (2) appear incorrectly in Morgan [25, Corollary 9] and Morgan and Johnson, respectively.

18.11 Perelman's Proof of the Poincaré Conjecture It has long been known that among connected compact two-dimensional manifolds, such as the sphere, the torus, and the two-holed torus, the sphere is characterized by the fact that any loop can be contracted to a point, whereas a loop around a torus, for example, cannot be contracted to a point. The Poincaré Conjecture, suggested by Henri Poincaré in 1904, proposes the analogous result for three-dimensional manifolds: *a simply connected compact three-dimensional manifold must be a sphere.* At the 2006 International Congress of Mathematicians, Grigori Perelman was awarded the Fields Medal for its proof, although he declined to accept it.

High-dimensional versions of the Poincaré Conjecture, with more space to do geometric constructions, are easier. Stephen Smale proved the analogous conjecture for dimensions at least 5 and won the Fields Medal in 1966. Michael Freedman proved the 4-dimensional case and won the Fields medal in 1986.

Figure 18.11.1. In 1904, Henri Poincaré (left) proposed his famous conjecture. In 2003, Grigori Perelman (right) proved it. [Photos from the Internet.]

The basic idea of Perelman's proof, due to Richard Hamilton, is to start with any simply connected compact three-manifold and let it shrink at each point in each direction at a rate proportional to its Ricci curvature. If you can show that you eventually end up with a round sphere, with perhaps other spheres pinched off along the way, you can conclude that you must have started with a (deformed) sphere.

The fundamental difficulty is to obtain some control over the formation of singularities. To focus attention about a point p of concern, Perelman [Sect. 1.1] gives the manifold large density about that point and lets the metric flow by the associated generalized Ricci curvature Ric_ψ 18.3(4). If one fixes the measure, then the density evolves as a modified backwards heat equation and approaches a delta function at p. His general monotonicity of energy now provides the requisite local information instead of the usual global information. (Actually, in the proof, to facilitate surgery, Perelman [Sect. 7] moves to a localized version of the density called the length function.)

Remarkably, modulo diffeomorphisms, this generalized Ricci flow is equivalent to the standard Ricci flow. As Perelman puts it, "The remarkable fact here is that different choices of [density] lead to the same flow, up to a diffeomorphism; that is, the choice of [density] is analogous to the choice of gauge." This means that during the proof one can choose any density for convenience. In summary, manifolds with density provide a convenient technical context for applying diffeomorphisms to focus attention on regions of concern.

In addition to general manifolds with density, the premier example of Gauss space makes an important appearance in Perelman's proof. Specifically, the isoperimetric inequality on Gauss space in functional form, known as the Gaussian logarithmic Sobolev inequality, is an important technical ingredient.

General manifolds with density have not appeared in most expositions of Perelman's work. This is partly because they appear just in his introductory sections and are replaced by his length function in the actual proof. It is also partly because manifolds with density are not as familiar to the mathematical community as they should be. They appear in the very first sentences of the body of Perelman's paper, where he begins by considering a manifold with density e^{-f} and scalar curvature R:

> Consider the functional $F = \int_M (R + |\nabla f|^2)e^{-f} dV$ for a Riemannian metric g_{ij} and a function f on a closed manifold M.

Anderson, in his *Notices* survey, makes the following comments on Perelman's discussion of Ricci flow in manifolds with density:

> It turns out that, given any initial metric $g(0)$ and $t > 0$, the [density] function $f \ldots$ can be freely specified at $g(t) \ldots$. Perelman then uses this freedom to probe the geometry of $g(t)$ with suitable choices of f. For instance, he shows by a very simple study of the form of [the energy] F that the collapse or noncollapse of the metric $g(t)$ near a point x_0 can be detected from the size of F by choosing f to be an approximation to

a delta function centered at x. The more collapsed $g(t)$ is near x, the more negative the value of F. The collapse of the metric $g(t)$ on any scale in finite time is then ruled out by combining this with the fact that the functional F is increasing along the Ricci flow. In fact, this argument is carried out with respect to a somewhat more complicated *scale-invariant* functional than F; motivated by certain analogies in statistical physics, Perelman calls this the entropy functional.

Perelman, in the following excerpt from his Introduction, refers to this noncollapse result of his §4. He also mentions the use in his §3 of the Gaussian isoperimetric inequality to prove that you do not get periodicity instead of the desired limiting behavior.

> We prove that Ricci flow, considered as a dynamical system on the space of Riemannian metrics modulo diffeomorphisms and scaling, has no nontrivial periodic orbits. The easy (and known) case of metrics with negative minimum of scalar curvature is treated in §2; the other case is dealt with in §3, using our main monotonicity formula (3.4) and the Gaussian logarithmic Sobolev inequality, due to L. Gross. In §4 we apply our monotonicity formula to prove that for a smooth solution on a finite time interval, the injectivity radius at each point is controlled by the curvatures at nearby points. This result removes the major stumbling block in Hamilton's approach to geometrization.

Exercises

18.1 Check that in the Gauss plane the unit circle has vanishing curvature. R. Bryant [Carroll *et al.*, Prop. A.1] proved that it is the only embedded closed geodesic.

18.2 Check that equality holds in the classical Heintze–Karcher theorem, 18.4, for the unit sphere about the origin in $\mathbf{R}^3$ with outward unit normal.

18.3 Check that strict inequality holds in the generalized Heintze–Karcher theorem, 18.5, for the unit sphere about the origin in $\mathbf{R}^3$ with outward unit normal.

18.4 Check that strict inequality holds in the classical Levy–Gromov theorem, 18.6, for a Euclidean 2-sphere of Ricci curvature $8 > \gamma = 2$, for the case of given volume 1/2.

18.5 Check that strict inequality holds in the generalized Levy–Gromov theorem, 18.7, for the unit Euclidean 2-sphere, for the case of given volume 1/2.

18.6 Check the variational formulae 18.9(1, 2) for a sphere in $\mathbf{R}^3$ with u the unit outward normal.

18.7 Check the variational formulae 18.9(1, 2) for a plane $\{x = t\}$ in 3D Gauss space G^3 with u the unit outward normal.

18.8 Check Corollary 18.10 for $\mathbf{R}^3$ and G^3.

18.9 Check that for $\mathbf{R}^3$ equality holds in the inequality 18.10(2).

Double Bubbles in Spheres, Gauss Space, and Tori

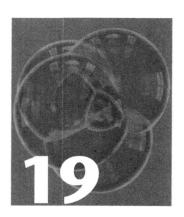

Following the proof by Hutchings *et al.* of the Double Bubble Conjecture in $\mathbf{R}^3$, there have been extensions to a few other spaces—the sphere, hyperbolic space, and Gauss space—as described here. In all of these cases, the minimizer is a canonical standard double bubble. In contrast, on a flat torus, there are many types of minimizers, already in dimension two. Even the circular two-dimensional cone has two types of minimizers by work of Lopez and Borawski (not presented here).

19.1 Double Bubbles in Spheres

It is conjectured that the least-perimeter way to enclose and separate two regions of prescribed volume in a round, n-dimensional sphere $\mathbf{S}^n$ is the *standard double bubble* (see Figure 14.0.1) consisting of three spherical caps meeting at 120 degrees. The proof [Cotton–Freeman, Prop. 2.6] of the existence of such a standard bubble, unique up to isometries, is similar to the proof in $\mathbf{R}^n$ (14.1). Since the sphere has finite volume, such a cluster is actually a partition of the sphere into three regions of prescribed volumes. The existence of a perimeter-minimizing double bubble in $\mathbf{S}^n$ follows immediately from the compactness theorem, 5.5, unlike the case of $\mathbf{R}^n$ (13.7), where there is the additional major concern that volume might in principle disappear to infinity in the limit. As in $\mathbf{R}^n$, a perimeter-minimizing double bubble is a hypersurface of revolution about a great circle, consisting of constant-mean-curvature surfaces meeting in threes along smooth curves at 120 degrees [Hutchings, Cotton–Freeman]. Unlike in $\mathbf{R}^n$, there are the additional possibilities that the cluster may be disjoint from or may totally envelop the axis of symmetry (see [Cotton–Freeman, Figure 6.3]).

Geometric Measure Theory. http://dx.doi.org/10.1016/B978-0-12-804489-6.00019-8

The main problem is that in principle each region may have many components. Fortunately, the Hutchings theory (Chapter 13) applies to S^n as well as R^n. In particular, there is the following bound on the number of components of a region. Let $A(v)$ denote the area of a hypersphere in S^n of volume v; let $A(v, w)$ denote the area of a perimeter-minimizing double bubble of volumes v, w.

19.2 Component Bound [Cotton–Freeman, Prop. 4.8] *Consider a perimeter-minimizing double bubble in S^n of volumes v, w. If for some integer $k \geq 2$, the Hutchings function*

(1) $$kA(v/k) + A(w) + A(v + w) - 2A(v, w)$$

is positive, then the region of volume v has fewer than k components. In particular, if

(2) $$2A(v/2) + A(w) + A(v + w) - 2A(v, w)$$

is positive, then the region of volume v is connected.

Remarks Formula (1) is the same as the R^n formula of 14.11, except without the scaling possible only in R^n. Of course, it suffices to show that (1) remains positive when $A(v, w)$ is replaced by the area of the standard double bubble.

In proofs of the Double Bubble Conjecture, there is a concluding instability argument to show that a perimeter-minimizing double bubble must be standard. The proof in R^3 (Chapter 14) showed this for bubbles of at most three components total (not counting the exterior). The generalization to R^4 (by Reichardt *et al.*) showed this for bubbles with one connected region. Reichardt's extension to R^n showed this without any bounds on the number of components. In the sphere, this has been shown only for all three regions (the two prescribed volumes and the exterior) connected.

19.3 Instability Argument [Cotton–Freeman, Prop. 7.3] *A perimeter-minimizing double bubble in S^n in which all three regions are connected must be the standard double bubble.*

19.4 Double Bubble Theorems in S^n Double bubble theorems in S^n to date have verified 19.2(2) in certain cases and then used 19.3 to conclude that a perimeter-minimizing double bubble must be standard. Unfortunately, hypothesis

19.2(2), although sufficient, is not necessary, and it fails in many cases. It is not known whether Reichard's component-bound-free methods for $\mathbf{R}^n$ (Section 14.0) can be generalized to $\mathbf{S}^n$.

For the exceptional case of $\mathbf{S}^2$, Masters (1996) computer checked a variant of 19.2(2) just for the largest region and then proved the Double Bubble Conjecture in $\mathbf{S}^2$ for all areas. For $\mathbf{S}^3$, Cotton and Freeman (2002) used computer analysis to verify 19.2(2) for all three regions and prove the Double Bubble Conjecture in $\mathbf{S}^3$ when the prescribed volumes are equal and the exterior occupies at least 10% of $\mathbf{S}^3$. Corneli, Hoffman, *et al.* (2007) used extensive and clever computer analysis to prove the Double Bubble Conjecture in $\mathbf{S}^3$ when each enclosed volume and the complement occupy at least 10% of the volume of $\mathbf{S}^3$. These results cover most of the territory where the hypothesis of 19.2 with $k = 2$ holds, and similar computer analysis in $\mathbf{S}^4$ seems impractical. The authors prove analogous results in hyperbolic space $\mathbf{H}^3$.

Finally, Corneli, Corwin, *et al.* (2007) proved the Double Bubble Conjecture in $\mathbf{S}^n$ when each region and the complement has volume fraction within .04 of 1/3. How did they verify 19.2(2) for all n in one fell swoop? They considered the Gauss plane G^2, the limit of orthogonal projections of $\mathbf{S}^n$, suitably normalized, as n approaches infinity (see Chapter 18). In Theorem 18.2, this relationship was used to solve the single bubble problem in G^2, to prove that lines bound regions (half-planes) of prescribed area and minimum perimeter in G^2. Here, one checks the Hutchings hypothesis 19.2(2) in G^2, whence it follows for $\mathbf{S}^n$ for large n and thence for all n. The asserted double bubble theorem for $\mathbf{S}^n$ follows.

The same relationship now yields results on G^n.

19.5   Double Bubble Theorem in G^n [Corneli, Corwin, *et al.*, Thm. 2.17]   *Three hyperplanes meeting at 120 degrees provide a perimeter-minimizing way to partition G^n into three regions of prescribed volumes each within .04 of 1/3.*

(Due to the variable density, translating this configuration yields all possible triples of volumes, uniquely up to isometry.)

Proof Sketch   As in the proof of Theorem 19.2, we use the fact that G^n is the limit of projections of high-dimensional spheres, suitably normalized. The triple hyperplane configuration is perimeter minimizing because its inverse projection, the spherical standard double bubble, is perimeter minimizing.

Corneli, Corwin, *et al.* [Conj. 2.20, Figure 8] conjecture that the perimeter-minimizing partition of G^2 into n areas consists of half-lines and line segments meeting in threes at 120 degrees.

19.6 Double Bubbles in Flat Two-Tori [Corneli, Holt, *et al.*] *On a flat two-torus, there are four or five types of perimeter-minimizing double bubbles as in Figures 19.6.1 and 19.6.2.*

Proof Outline Existence follows from the compactness theorem, 5.1. Regularity (see Section 13.10) states that minimizers consist of finitely many circular arcs meeting in threes at 120 degrees.

The proof begins by classifying five possible combinatorial types: (1) contractible clusters, (2) the cluster but neither region wraps around the torus, (3) one region wraps around the torus, (4) both regions wrap around the torus, and (5) tilings. In the first type, since a standard double bubble is the minimizer in the plane, it will remain the minimizer on the torus, if it fits. That it does fit follows from the lemma that the perimeter P and diameter D of a standard double bubble satisfy the inequality [Corneli, Holt, *et al.*, Prop. 5.1.1].

(1) $$P > \pi D,$$

(with equality in the limiting case of a single bubble).

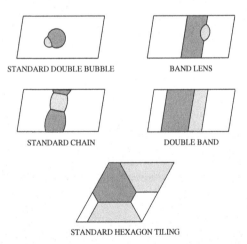

STANDARD DOUBLE BUBBLE   BAND LENS

STANDARD CHAIN   DOUBLE BAND

STANDARD HEXAGON TILING

Figure 19.6.1. On a flat two-torus, there are four or five types of perimeter-minimizing double bubbles. Flat tori are canonically represented by parallelograms with small angle ranging from $\pi/3$ to $\pi/2$, sides no longer than top and bottom, and opposite edges identified. The hexagon tiling, in which both regions and the complement are hexagons, occurs only for the "hexagonal torus," a rhombus with angle $\pi/3$, where it always ties the double band. From Corneli, Holt, *et al.*, Figure 1, used by permission.

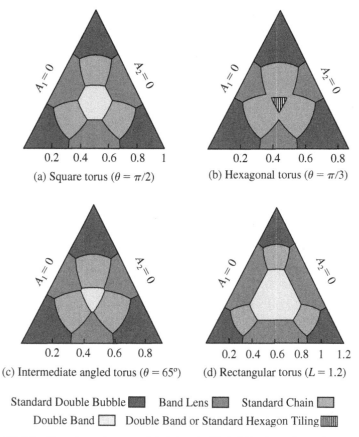

(a) Square torus ($\theta = \pi/2$)   (b) Hexagonal torus ($\theta = \pi/3$)

(c) Intermediate angled torus ($\theta = 65°$)   (d) Rectangular torus ($L = 1.2$)

Standard Double Bubble ▉   Band Lens ▨   Standard Chain ▢
Double Band ▢   Double Band or Standard Hexagon Tiling ▥

Figure 19.6.2. How the perimeter-minimizing double bubble depends on the prescribed areas for four specific two-tori. In each triangle, the first prescribed area vanishes along the left edge, the second along the right edge, and the complementary area along the bottom. Down the middle, the two prescribed areas are equal. The hexagon tiling, in which both regions and the complement are hexagons, occurs only for the "hexagonal torus," a rhombus with angle $\pi/3$, where it always ties the double band. From Corneli, Holt, *et al.*, Figure 2, used by permission.

In the second type, which looks in general like Figure 19.6.3, appendages are ruled out by showing how reflection of an outermost appendage and an adjacent edge would yield an illegal singularity.

The rest of the long proof uses many such geometric ideas as well as delicate estimates and computations. The most difficult case to eliminate was the notorious

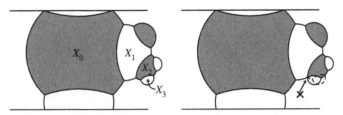

Figure 19.6.3. Among chains, appendages are ruled out by showing how reflection of an outermost appendage X_3 and an adjacent edge would yield an illegal singularity. From Corneli, Holt, *et al.*, Figure 10, used by permission.

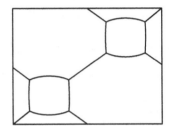

Figure 19.6.4. The most difficult case to eliminate was the notorious "octagon–square" tiling, in which one region consists of two curvilinear squares, whereas the other two regions are each octagonal. From Corneli, Holt, *et al.*, Figure 23, used by permission.

"octagon–square" tiling of Figure 19.6.4, in which one region consists of two curvilinear squares, whereas the other two regions are each octagonal.

19.7 The Cubic Three-Torus
For the flat cubic three-torus, even the best single bubbles are conjectural: balls, cylinders, slabs, and complements (Section 13.2). Numerical studies by Carrión *et al.* on Brakke's Surface Evolver indicate that there are 10 different types of perimeter minimizers, as shown in Figures 19.7.1–19.7.4.

Exercises

19.1 For Figure 19.6.2(a), compute exactly where the regions meet along the boundary of the triangle.

19.2 Prove inequality 19.6(1).

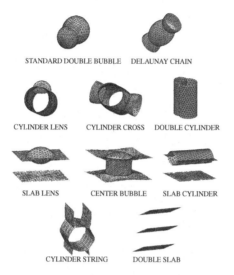

Figure 19.7.1. The 10 conjectured types of perimeter-minimizing double bubbles in the cubic three-torus. From Carrión *et al.*, Figure 1, used by permission.

Figure 19.7.2. Brubaker *et al.* physically obtained the slab cylinder, the center bubble, and others not shown here.

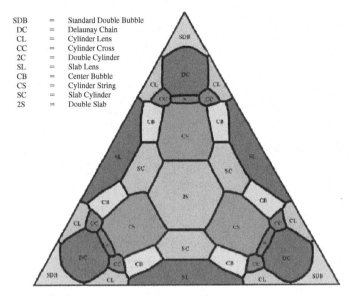

SDB = Standard Double Bubble
DC = Delaunay Chain
CL = Cylinder Lens
CC = Cylinder Cross
2C = Double Cylinder
SL = Slab Lens
CB = Center Bubble
CS = Cylinder String
SC = Slab Cylinder
2S = Double Slab

Figure 19.7.3. How the perimeter-minimizing double bubble depends on the prescribed volumes. In the center, both regions and the complement have one-third of the total volume; along the edges one volume is small; in the corners two volumes are small. From Carrión *et al.*, Figure 2, used by permission.

Figure 19.7.4. Carrión *et al.* commissioned a stained-glass window of Figure 19.7.3, which the author proudly displays in his office window.

The Log-Convex Density Theorem

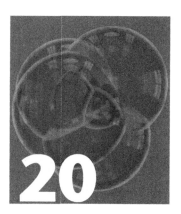

Since their appearance in Perelman's 2003 proof of the 1904 Poincaré Conjecture, there has been a huge surge of interest in placing a density or weight on a space [Morgan 29] and considering the isoperimetric problem of finding the least weighted perimeter to enclose given weighted volume. Even if the density is radial, balls about the origin may not be isoperimetric. Indeed, as described in Chapter 18, for Gaussian density isoperimetric regions are halfspaces.

The Log-Convex Density Theorem, finally proved by Gregory Chambers in 2014, says that balls about the origin are isoperimetric for a smooth radial density if and only if the log of the density is convex. First we show that the condition is necessary because it is precisely the condition for stability (nonnegative second variation). We begin with the second variation formula due to Vincent Bayle (see [Rosales *et al.*, Prop. 3.6]). It applies to hypersurfaces for which the first variation of area vanishes or, equivalently, for which the generalized mean curvature H_ψ is constant (see §18.3 and 18.9(1)).

20.1 Proposition *Consider $\mathbf{R}^n$ with C^2 density $\Psi = e^\psi$. Let R be a smooth, compact region. Suppose that the first variation of the area of ∂R vanishes for fixed volume (H_ψ constant). Then the second variation of perimeter for normal velocity u preserving volume to first order satisfies*

$$(1) \qquad P'' = \int_{\partial R} |\nabla_{\partial R} u|^2 - \mathrm{II}^2 u^2 + u^2 (\partial^2 \psi / \partial n^2).$$

Here II^2 is the square of the second fundamental form, the sum of the squares of the principal curvatures, and the integral is with respect to weighted area.

Geometric Measure Theory. http://dx.doi.org/10.1016/B978-0-12-804489-6.00020-4

Proof   Since the first variation vanishes, the second variation depends only on the initial velocity, and every initial velocity preserving weighted volume to first order ($\int u = 0$) corresponds to some volume-preserving variation. For convenience we consider variations of constant velocity along geodesics normal to ∂R. Then the second variation of volume is due to change in area (see 18.9(1)) and is given by

$$V'' = - \int u^2(n-1)H_\psi.$$

Consequently by 18.9(2) with Ric = 0,

$$(P - H_\psi V)'' = \int |\nabla u|^2 - \mathrm{II}^2 u^2 + u^2(\partial^2 \psi/\partial \mathrm{n}^2).$$

In particular, with V constant, the proposition follows.   ∎

20.2   Remarks   In a smooth Riemannian manifold with density the second variation has an additional term depending on the Ricci curvature of the manifold in the normal direction, $- \int u^2\mathrm{Ric}$. Consequently 20.1(1), in terms of the generalized Ricci curvature Ric_ψ in the normal direction (see 18.7, 18.9), becomes

$$P'' = \int |\nabla u|^2 - \mathrm{II}^2 u^2 - u^2\mathrm{Ric}_\psi.$$

More generally, one can consider different densities f and g for volume and perimeter (see Howe and Morgan [30]). We've been considering the case $f = g$. A conformal change of metric corresponds to $f^{n-1} = g^n$. The situation is simpler for volume density, where the stability condition (volume density nonincreasing) trivially implies that balls about the origin are isoperimetric (Exercise 20.8). Morgan [31] provides the general second variation formula.

The consequence of the second variation formula 20.1(1) for balls about the origin, first computed at a conference in 2004 by Ken Brakke, gives the following simple criterion for stability:

20.3   Corollary [Rosales *et al.*, Thm. 3.10]   *Consider* $\mathbf{R}^n$ *with* C^2 *density* $\Psi = e^\psi$. *Then balls about the origin have nonnegative second variation for fixed volume if and only if* $\psi''(\mathrm{r}) \geq 0$ *for all* r, *that is, if and only if the density is log convex.*

Proof   By symmetry balls about the origin have vanishing first variation. Hence second variation depends only on the initial velocity. Moreover, every initial velocity preserving volume to first order corresponds to a variation preserving velocity. So we need consider the second variation formula 20.1(1) precisely for (smooth) initial normal velocities u with mean 0. Since Euclidean balls have nonnegative second variation in $\mathbf{R}^n$ with the standard density $\Psi = 1$, the integral of the first two terms is nonnegative and vanishes for translations. Consequently if in the third term $\partial^2 \psi / \partial n^2 = \psi''(r) \geq 0$, the second variation is nonnegative. Conversely, if the second variation is nonnegative for translations, since ψ is constant on a sphere about the origin, $\psi''(r) \geq 0$. Finally note that since the Hessian of ψ'' has eigenvalues proportional to $\psi''(r)$ and $\psi'(r)$, $\psi''(r)$ is convex if and only if $\psi''(r) \geq 0$ (which implies that $\psi'(r) \geq 0$). ∎

20.4 Existence and regularity
A key ingredient in the proof of the Log-Convex Density Conjecture is the existence of isoperimetric regions, which fails for some radial densities on $\mathbf{R}^n$, such as r^{-1} on $\mathbf{R}^2$, for which expanding annuli of fixed weighted area have perimeter approaching 0. In general isoperimetric regions prefer low density, because when you scale them up to have the prescribed weighted volume, the weighted perimeter scales at a lower rate. In 2013 Morgan and Pratelli proved that for a C^1 radial density diverging to infinity on $\mathbf{R}^n$, including all nontrivial C^1 log-convex radial densities, isoperimetric regions exist for all prescribed volumes, with later improvements by De Philippis, Franzina, and Pratelli. One considers a sequence of regions of the prescribed (weighted) volume with (weighted) perimeter approaching the infimum. As in Section 13.7, there is an isoperimetric limit, but some part of the volume may have disappeared to infinity. In that case, one shows that either the radial or the tangential contributions to the perimeter are large, the desired contradiction.

As mentioned in Section 8.5, an isoperimetric hypersurface is regular except for a singular set of Hausdorff codimension at least 7.

20.5 History of the Log-Convex Density Conjecture
In my lectures on "Geometric Measure Theory and Isoperimetric Problems" at the 2004 Summer School on Minimal Surfaces and Variational Problems held in the Institut de Mathématiques de Jussieu in Paris, organized by Pascal Romon, Marc Soret, Rabah Souam, Eric Toubiana, Frédéric Hélein, David Hoffman, Antonio Ros and Harold Rosenberg, I naively challenged the students to prove that for an increasing, radial density of $\mathbf{R}^n$, balls about the origin are isoperimetric. Overnight Ken Brakke computed the stability criterion for nonnegative second variation, namely that the log of the density be convex, showing the silliness of my challenge and producing the new conjecture that his necessary condition was sufficient.

After the trivial, borderline case of Euclidean space with unit density, the second case that had been proved was density $\Psi(r) = \exp(r^2)$, by Christer Borell [Rosales *et al.*, Sect. 5]. The proof used Steiner symmetrization, a method useful only for multiples of $\exp(cr^2)$, the only rotationally symmetric densities which are also products; for example, in the plane, $\exp(r^2) = \exp(x^2) \cdot \exp(y^2)$. (See Exercise 20.4.) Symmetrization fails for non-product densities [Betta *et al.*, Thm. 3.10].

The third example was $\exp(r^p)$ for $p \geq 2$ in $\mathbf{R}^2$ by Maurmann and Morgan, by comparison with certain classical surfaces of revolution. The cases $1 \leq p < 2$ remained conjectural.

The fourth example was large balls in $\mathbf{R}^n$ with uniformly log-convex density (Kolesnikov and Zhdanov, generalized by Howe), by the Divergence Theorem.

The fifth example was small balls in $\mathbf{R}^n$ with uniformly log-convex density by asymptotic analysis by Allesio Figalli and Francesco Maggi.

There were also some fascinating examples in convex cones; see [Morgan, 29].

In 2013 the result was proved by Greogory Chambers, a finishing graduate student at the University of Toronto:

20.6 The Log-Convex Density Theorem (Chambers, 2013)

Consider $\mathbf{R}^n$ with smooth, radial density. Then balls about the origin are isoperimetric if and only if the density is log convex, with uniqueness unless the density is constant in a neighborhood of the origin.

Outline of proof Start with any candidate. Apply spherical symmetrization, replacing its intersection with every sphere about the origin with a polar cap of the same area. Since symmetrization preserves volume but generally reduces perimeter, one may assume that the candidate is a solid of revolution about the polar axis, say the e_1 axis, hence determined by its generating region in the e_1e_2-plane. Since the area of the candidate has vanishing first variation for given volume, the bounding curve in the plane satisfies a certain differential equation. Chambers studies such curves very intelligently in great detail. The main idea is that if the generating curve is not a circle about the origin, then as in Figure 20.5.1, from its maximum (on the axis of symmetry) it spirals inward and it eventually turns through 2π before returning to the axis, contradiction. The obvious estimates, however, do not hold, so the analysis is subtle and involves comparing points on the curve at equal heights. See the 30-second video of Chambers on YouTube. ∎

Extensions The conjecture has natural open extensions to hyperbolic space, to the sphere, and to more general surfaces of revolution, where the conjecture would be that if balls about the pole are isoperimetric with density 1, then they are isoperimetric for any log-convex radial density.

Generalizing Chambers's methods, Morgan's NSF "SMALL" undergraduate research 2014 Geometry Group (Figure 20.7.2), led by Sarah Tammen, working with Chambers, was able to prove a more surprising conjecture about another radial density, any positive power r^p of the distance from the origin. The planar case had been proved by the 2008 Geometry Group [Dalberg *et al.*].

20.7 Theorem (Boyer *et al.*) *Consider $\mathbf{R}^n$ with density r^p, $p > 0$. Then spheres through the origin are uniquely isoperimetric.*

Note that the solution does not have the symmetry of the problem, so that there are infinitely many such spheres for each prescribed volume.

Remarks on Proof Although the proof follows the same line as the Log-Convex Density Conjecture's, there are additional difficulties. In the latter, the presumptive better generating curve can only have *greater* curvature than the circle at the rightmost point farthest from the origin, as in Figure 20.6.1. In the former, there are two difficult cases, pictured in Figure 20.7.1.

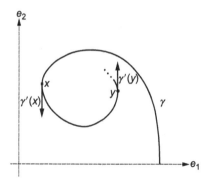

Figure 20.6.1. If the generating curve is not a circle about the origin, it spirals inward. [Chambers, Figure 2]. Permission from Gregory Chambers.

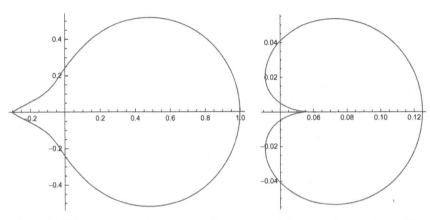

Figure 20.7.1. The generating curve at the extreme, rightmost point can have either too little curvature or too much curvature, both leading to contradictions [Boyer *et al.*, Figure 2].

Figure 20.7.2. Morgan's 2014 NSF "SMALL" undergraduate research Geometry Group with Greg Chambers: Sarah Tammen, Bryan Brown, Alyssa Loving, Chambers, Wyatt Boyer.

Exercises

20.1 Use 20.1(1) to compute that under horizontal translation, the unit circle in the plane with smooth log-convex density has nonnegative second variation.

20.2 Why isn't density r^{-1} on the plane a counterexample to the Log-Convex Density Theorem 20.6?

20.3 Prove that smooth density on $\mathbf{R}^n$ is radial if and only if spheres about the origin are stationary for given volume [Li *et al.*, Prop. 6.2].

20.4 Prove that a smooth radial product density on $\mathbf{R}^2$ must be of the form $a \exp(br^2)$.

20.5 Give an example of a smooth surface of revolution (unit density) for which balls about the origin are not isoperimetric.

20.6 Compute that in the plane with density $e^{\psi} = r^p (p > 0)$, circles through the origin have constant generalized curvature $\kappa - d\psi/dn$.

20.7 Use 20.1(1) to compute that in $\mathbf{R}^n$ with density r^2, balls about the origin do not have nonnegative second variation for fixed volume.

20.8 Prove directly that for unit perimeter density and nondecreasing radial volume density on $\mathbf{R}^n$, balls about the origin are isoperimetric.

Solutions to Exercises

Chapter 2

2.0. Such a scaling follows quickly from the definition. To prove that $H^m(rA)$ $\leq r^m H^m(A)$, note that if $\cup S_j$ with $\mathrm{diam}(S_j) \leq \delta$ cover A, then $\cup(rS_j)$ with $\mathrm{diam}(rS_j) \leq r\delta$ cover rA. The proof of the opposite inequality is similar.

2.1.

$$\mathscr{I}^1(I) \equiv \frac{1}{\beta(2,1)} \int_{p \in 0^*(2,1)} \int_{y \in \mathrm{Im} p} N(p|I, y) d\mathscr{L}^1 dp$$

$$= \frac{\pi}{2} \int_0^\pi |\cos\theta| \frac{d\theta}{\pi} = 1.$$

2.2. Coverings by n intervals of length $1/n$ show that $\mathscr{H}^1(I) \leq 1$. Suppose $\mathscr{H}^1(I) < 1$. Then there is a covering $\{S_j\}$ of I with

$$\sum \mathrm{diam}\, S_j < 1.$$

By slightly increasing each $\mathrm{diam}\, S_j$ if necessary, we may assume that the S_j are open intervals (a_j, b_j). Since I is compact, we may assume that there are only finitely many. We may assume that none contains another. Finally, we may assume that $a_1 < a_2 < \cdots < a_n$ and hence $b_j > a_{j+1}$. Now

$$\sum_{j=1}^n \mathrm{diam}\, S_j = \sum_{j=1}^n (b_j - a_j) \geq \sum_{j=1}^{n-1} (a_{j+1} - a_j) + (b_n - a_n)$$

$$= b_n - a_1 > 1,$$

the desired contradiction.

2.3. Covering $[-1, 1]^n$ by $(2N)^n$ cubes of side $1/N$ and radius $\sqrt{n}/2N$ yields

$$\mathscr{H}^n(\mathbf{B}^n(0, 1)) \leq \mathscr{H}^n([-1, 1]^n) \leq \lim(2N)^n \alpha_n (\sqrt{n}/2N)^n = \alpha_n n^{n/2} < \infty.$$

2.4. For each $\delta > 0$, there is a cover $\{S_j(\delta)\}$ of A with $\text{diam}(S_j(\delta)) \leq \delta$ and

$$\sum \alpha_m \left(\frac{\text{diam } S_j(\delta)}{2} \right)^m \leq \mathcal{H}^m(A) + \varepsilon < \infty.$$

Consequently,

$$\lim \sum \alpha_k \left(\frac{\text{diam } S_j(\delta)}{2} \right)^K$$

$$= \frac{\alpha_k}{\alpha_m} \lim \sum \alpha_m \left(\frac{\text{diam } S_j(\delta)}{2} \right)^m \left(\frac{\text{diam } S_j(\delta)}{2} \right)^{k-m}$$

$$\leq \frac{\alpha_k}{\alpha_m} (\mathcal{H}^m(A) + \varepsilon) \lim \left(\frac{\delta}{2} \right)^{k-m} = 0.$$

Therefore, $\mathcal{H}^k(A) = 0$.

It follows that for a fixed set A, there is a nonnegative number d such that

$$\mathcal{H}^m(A) = \begin{cases} \infty & \text{if } 0 \leq m < d, \\ 0 & \text{if } d < m < \infty. \end{cases}$$

All four definitions of the Hausdorff dimension of A yield d. Incidentally, $\mathcal{H}^d(A)$ could be anything: 0, ∞, or any positive real number, depending on what A is.

2.5. The 3^j triangular regions of side 3^{-j} that make up A_j provide a covering of A with $\Sigma \alpha_1 (\text{diam}/2)^1 = 1$. Hence, $\mathcal{H}^1(A) \leq 1$.

The opposite inequality is usually difficult, but here there is an easy way. Let Π denote projection onto the x-axis. Then $\mathcal{H}^1(A) \geq \mathcal{H}^1 (\Pi(A)) = 1$.

2.6. (a) C can be covered by 2^n intervals of length 3^{-n}.

(b) Suppose $\mathcal{H}^m(C) < \alpha_m/2^m$, so there is a covering by intervals S_i with $\Sigma(\text{diam } S_i)^m < 1$. By slight enlargement, we may assume for the moment that each S_i is open. Since C is compact, we may assume that $\{S_i\}$ is finite, of minimal cardinality, and, now taking the S_i closed, for that cardinality with $\sum (\text{diam} S_i)^m$ as small as possible. If no S_i meets both halves of C, a scaling up of the covering of one of the two halves yields a cheap covering of C of smaller cardinality, a contradiction. If exactly one S_i meets both halves of C, replace it by

$$\{x \in S_i : x \leq \tfrac{1}{3}\} \text{ and } \{x \in S_i : x \geq \tfrac{2}{3}\}.$$

Now a scaling up of the covering of the cheaper half yields a strictly cheaper covering of C of no larger cardinality, a contradiction. If

$p \geq 2$ S_i's meet both halves, we may suppose that the leftmost is S_1 and the rightmost S_2. Replace all p of them with

$$S_1' = \{x \in X_1 : x \leq \tfrac{1}{3}\} \text{ and } S_2' = \{x \in S_2 : x \geq \tfrac{2}{3}\}$$

to obtain a strictly cheaper covering with no greater cardinality and the desired contradiction.

Thanks to G. Lawlor for finding the error in the solution in the second and third editions, which wrongly assumed that the cheap half also had small cardinality.

2.7.

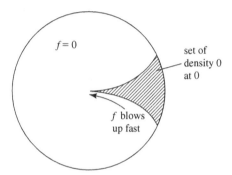

2.8. Let $\varepsilon > 0$. Suppose **0** is a Lebesgue point of f. Then

$$\frac{1}{\alpha_m r^m} \int_{\mathbf{B}^m(0,r)} |f(x) - f(0)| d\mathscr{L}^m \to 0.$$

Consequently,

$$\frac{1}{\alpha_m r^m} \mathscr{L}^m \{x \in \mathbf{B}^m(0,r) : |f(x) - f(0)| \geq \varepsilon\} \to 0.$$

Therefore, f is approximately continuous at **0**.

2.9. Following the hint, let $a \in \cap E_i$. It suffices to show that f is approximately continuous at a since by Corollary 2.9, almost every point lies in $\cap E_i$.

Given $\varepsilon > 0$, choose

$$f(a) - \varepsilon < q_i < f(a) < q_j < f(a) + \varepsilon.$$

Then

$$\Theta(\{|f(x) - f(a)| \geq \varepsilon\}, a)$$
$$\leq \Theta(\{f(x) < q_i\}, a) + \Theta(\{f(x) > q_j\}, a) = 0 + 0$$

because $a \in E_i$ and $a \in E_j$.

Chapter 3

3.1. Let $\{q_i\}$ be an enumeration of the rationals and let

$$f(x) = \sum_{i=1}^{\infty} 2^{-i}|x - q_i|.$$

3.2. Given $\varepsilon > 0$, f is approximately differentiable at the points of density 1 of $\{x \in A : f(x) = g(x)\}$—that is, everywhere except for a set of measure $< \varepsilon$.

3.3. On all of **R**, one can just take $f(x) = x^2$. On $[-1, 1]$, one can take $f(x) = \sqrt[3]{x}$.

3.4.

$$J_1 f = \begin{cases} 1 & \text{for } r \geq 1, \\ 1/r & \text{for } r \leq 1. \end{cases}$$

3.5.

$$J_1 f \equiv \max\{Df(u) : u \text{ unit vector}\}$$
$$= \max\{\nabla f \cdot u\} = |\nabla f|.$$

3.6.

$$\mathscr{H}^2(\mathbf{S}^2(\mathbf{0}, 1)) = \int_0^{2\pi} \int_0^{\pi} J_2 f \, d\varphi d\theta.$$
$$J_2 f = \sin \varphi.$$

The answer is 4π.

3.7.

$$\text{LHS} = \int_A J_1 f \, d\mathscr{L}^3 = \int_{\mathbf{B}(\mathbf{0}, R)} 2r \, d\mathscr{L}^3 = \int_0^R (2r)(4\pi r^2) \, dr = 2\pi R^4.$$

$$\text{RHS} = \int_0^{R^2} 4\pi y \, dy = 2\pi R^4.$$

3.8. (a) Apply the area formula, 3.13, to $f : E \times [0, 1] \to C$ given by $f(x, t) = xt$.

$$\mathcal{H}^{m+1}(C) = \int_{E \times [0,1]} J_{m+1} f = \int_{E \times [0,1]} t^m = \frac{a_0}{m+1}.$$

Alternatively, apply the coarea formula, 3.13, to $f : C \to \mathbf{R}$ given by $f(x) = |x|$. Then

$$\mathcal{H}^{m+1}(C) = \int_0^1 a_0 y^m = \frac{a_0}{m+1}.$$

(b)

$$\Theta^{m+1}(C, \mathbf{0}) = \lim \frac{\mathcal{H}^{m+1}(C \cap B(\mathbf{0}, r))}{\alpha_{m+1} r^{m+1}}$$

$$= \lim \frac{a_0 r^{m+1}/(m+1)}{\alpha_{m+1} r^{m+1}}$$

$$= \frac{a_0}{\alpha_{m+1}(m+1)}.$$

(c) The cone over $\{x \in E : \mathcal{H}^m(B(x, r) \cap E) > 0 \text{ for all } r > 0\}$.

3.9. Let $\{q_i\}$ be an enumeration of the points in $\mathbf{R}^3$ with rational coordinates. Let $E = \cup_{i=1}^\infty S(q_i, 2^{-i})$. By 3.12, $\Theta^2(E, x) = 1$ for almost all $x \in E$. It follows that $\{x \in \mathbf{R}^3 : \Theta^2(E, x) = 1\}$ is dense in $\mathbf{R}^3$.

3.10. It follows from the triangle inequality that f is Lipschitz with Lipschitz constant 1. Suppose that f is differentiable. For any nearest point a of A, the derivative in that direction is -1, so that grad f must point directly away from a, and a must be unique.

Conversely, suppose that a is unique. We may assume that x_0 is the origin of $\mathbf{R}^n$ and that $a = (0, \ldots 0, -1)$. We want to prove that near 0, $f(x) = 1 + x_n + o(x)$, where $x_n = |x| \cos \theta$. Consideration of the point a shows that $f(x) \le 1 + x_n + o(x)$. Given small $\varepsilon > 0$, choose smaller $\delta > 0$ such that $|x| < \delta$ implies that a nearest point b of A is within ε of a as well as outside the unit ball. It follows that $f(x) \ge 1 + x_n + o(x)$.

Chapter 4

4.1. $-6\mathbf{e}_{134} - 12\mathbf{e}_{234}$.

4.2. One possibility is

$$u = (-1, 0, 1, -1),$$

$$v = (0, -1, 1, -1),$$

$$w = \left(-\frac{1}{\sqrt{3}}, 0, \frac{1}{\sqrt{3}}, -\frac{1}{\sqrt{3}}\right),$$

$$z = \left(\frac{2}{\sqrt{15}}, -\frac{3}{\sqrt{15}}, \frac{1}{\sqrt{15}}, -\frac{1}{\sqrt{15}}\right),$$

$$u \wedge v = \mathbf{e}_{12} - \mathbf{e}_{13} + \mathbf{e}_{14} + \mathbf{e}_{23} - \mathbf{e}_{24},$$

$$w \wedge z = \frac{1}{\sqrt{5}}\mathbf{e}_{12} - \frac{1}{\sqrt{5}}\mathbf{e}_{13} + \frac{1}{\sqrt{5}}\mathbf{e}_{14} + \frac{1}{\sqrt{5}}\mathbf{e}_{23} - \frac{1}{\sqrt{5}}\mathbf{e}_{24}$$

$$= \frac{1}{\sqrt{5}}u \wedge v,$$

$$|w \wedge z| = 1.$$

4.4. $\mathbf{e}_{12} + 2\mathbf{e}_{13} + 2\mathbf{e}_{23} = (\mathbf{e}_1 + \mathbf{e}_2) \wedge (\mathbf{e}_2 + 2\mathbf{e}_3)$.

4.5. Method 1: Assume $\mathbf{e}_{12} + \mathbf{e}_{34} = (\Sigma a_i \mathbf{e}_i) \wedge (\Sigma b_j \mathbf{e}_j)$, and derive a contradiction.

Method 2: Clearly, if ξ is simple, $\xi \wedge \xi = 0$. Since $(\mathbf{e}_{12} + \mathbf{e}_{34}) \wedge (\mathbf{e}_{12} + \mathbf{e}_{34}) = 2\mathbf{e}_{1234} \neq 0$, it is not simple. (Actually, for $\xi \in \Lambda_2 \mathbf{R}^n$, ξ simple $\Leftrightarrow \xi \wedge \xi = 0$. For $\xi \in \Lambda_m \mathbf{R}^n$, $m > 2$, ξ simple $\Rightarrow \xi \wedge \xi = 0$.)

4.6. $\int_0^1 \int_0^1 \langle \mathbf{e}_{12}, \varphi \rangle dx_2\, dx_1$, $\langle \mathbf{e}_{12}, \varphi \rangle = x_1 \sin x_1 x_2$.
Inside integral $= -\cos x_1 x_2]_{x_2=0}^1 = 1 - \cos x_1$.
Outside integral $= x_1 - \sin x_1]_{x_1=0}^1 = 1 - \sin 1$.

4.7. The surface is a unit disc with normal $\mathbf{e}_1 + \mathbf{e}_2 + \mathbf{e}_3$, unit tangent $\xi = (\mathbf{e}_{12} + \mathbf{e}_{23} - \mathbf{e}_{13})/\sqrt{3}$. $\varphi(\xi) = 4/\sqrt{3}$. Integral $= 4\pi\sqrt{3}$.

4.8. If $I \in \mathbf{I}_m$, then $\partial I \in \mathscr{R}_{m-1}$ by definition and $\partial(\partial I) = 0 \in \mathscr{R}_{m-2}$. Therefore, $\partial I \in \mathbf{I}_{m-1}$. If $F \in \mathscr{F}_m$, then $F = T + \partial S$, with $T \in \mathscr{R}_m$, $S \in \mathscr{R}_{m+1}$. Since $\partial F = \partial T$, $\partial F \in \mathscr{F}_{m-1}$.

To prove that spt $\partial T \subset$ spt T, consider a form $\varphi \in \mathscr{D}^{m-1}$ such that spt $\varphi \cap$ spt $T = \varnothing$. Then spt $d\varphi \cap$ spt $T = \varnothing$, and consequently $\partial T(\varphi) = T(d\varphi) = 0$. We conclude that spt $\partial T \subset$ spt T.

4.9. (a)

$$T(f\,dx + g\,dy) = \int_0^1 f(x,0)\,dx.$$

$$\partial T(h) = T\left[\frac{\partial h}{\partial x}dx + \frac{\partial h}{\partial y}dy\right]$$

$$= \int_0^1 \frac{\partial h}{\partial x}(x,0)\,dx = h(1,0) - h(0,0).$$

Hence, $\partial T = \mathscr{H}^0\llcorner\{(1,0)\} - \mathscr{H}^0\llcorner\{(0,0)\}$.

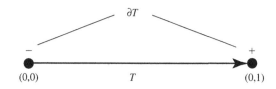

(b) Let $E = \{(x,x) : 0 \le x \le 1\}$.

$$T(f\,dx + g\,dy) = \int_E 3\sqrt{2}(f(t,t) + g(t,t))\,d\,\mathscr{H}^1$$

$$= \int_0^1 3\sqrt{2}(f(t,t) + g(t,t))\sqrt{2}dt$$

$$= 6\int_0^1 (f(t,t) + g(t,t))\,dt.$$

$$\partial T(h) = 6\int_0^1 \left[\frac{\partial h}{\partial x}(t,t) + \frac{\partial h}{\partial y}(t,t)\right]dt$$

$$= 6(h(1,1) - h(0,0)).$$

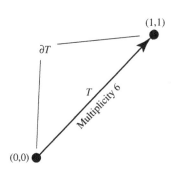

Notice that

$$T = 6 \mathcal{H}^1 \lfloor E \wedge \frac{\mathbf{e}_1 + \mathbf{e}_2}{\sqrt{2}}.$$

4.10. It follows from Theorem 4.4(1) that $\mathbf{I}_m$ is $\mathbf{M}$ dense in $\mathscr{R}_m$. To show that $\mathbf{I}_m$ is $\mathscr{F}$ dense in $\mathscr{F}_m$, let $R \in \mathscr{F}_m$ so that $R = T + \partial S$, with $T \in \mathscr{R}_m$ and $S \in \mathscr{R}_{m+1}$. Given $\varepsilon > 0$, choose $T_1 \in \mathbf{I}_m$, $S_1 \in \mathbf{I}_{m+1}$, such that $\mathbf{M}(T_1 - T) + \mathbf{M}(S_1 - S) < \varepsilon$. Then $T_1 + \partial S_1 \in \mathbf{I}_m$, and $\mathscr{F}((T_1 + \partial S_1) - R) = \mathscr{F}((T_1 - T) + \partial(S_1 - S)) \le \mathbf{M}(T_1 - T) + \mathbf{M}(S_1 - S) < \varepsilon$.

4.11. It follows from Theorem 4.4(1) that $A = \{T \in \mathscr{R}_m : \text{spt } T \subset \mathbf{B}(0, R)\}$ is $\mathbf{M}$ complete. To show that $B = \{T \in \mathscr{F}_m : \text{spt } T \subset \mathbf{B}(0, R)\}$ is $\mathscr{F}$ complete, let R_j be a Cauchy sequence in B. By taking a subsequence, we may assume that $\mathscr{F}(R_{j+1} - R_j) < 2^{-j}$. Write $R_{j+1} - R_j = T_j + \partial S_j$ with $\mathbf{M}(T_j) + \mathbf{M}(S_j) < 2^{-j}$. Since A is $\mathbf{M}$ complete, ΣT_j converges to a rectifiable current T, and ΣS_j converges to a rectifiable current S. It is easy to check that $R_j \to R_1 + T + \partial S \in \mathscr{F}_m$.

4.12. That ∂ carries $\mathbf{N}_m$ into $\mathbf{N}_{m-1}$ follows immediately from the definition of $\mathbf{N}_m$ and the fact that $\partial\partial = 0$. Since ∂ is $\mathbf{F}$ continuous, it follows that ∂ carries $\mathbf{F}_m$ into $\mathbf{F}_{m-1}$.

4.13. The first follows immediately from the definitions. The second is the definition of $\mathbf{R}_m$.

4.14. The first follows immediately from the definitions. By Exercise 4.12, $\mathbf{F}_m \supset \{T + \partial S : T \in \mathbf{R}_m, S \in \mathbf{R}_{m+1}\}$. Conversely, suppose $R \in \mathbf{F}_m$. Let $N_j \in \mathbf{N}_m$ with $\mathbf{F}(N_j - R) < 2^{-j-1}$ and hence $\mathbf{F}(N_{j+1} - N_j) < 2^{-j}$. Hence, $(N_{j+1} - N_j) = A_j + \partial B_j$ for currents A_j and B_j with $\mathbf{M}(A_j) + \mathbf{M}(B_j) < 2^{-j}$. Since $\mathbf{M}(\partial B_j) = \mathbf{M}(N_{j+1} - N_j - A_j) < \infty$, $B_j \in \mathbf{N}_{m+1}$. Therefore, $A_j = N_{j+1} - N - \partial B_j \in \mathbf{N}_m$. By Proposition 4.6, $\Sigma A_j \in \mathbf{R}_m$ and $\Sigma B_j \in \mathbf{R}_{m+1}$. Finally, $R = (N_1 + \Sigma A_j) + \partial \Sigma B_j$ is of the form $T + \partial S$, as desired.

4.15. That $\mathbf{I}_m \subset \mathbf{N}_m$ follows immediately from the definitions. Next,

$$\mathscr{F}_m \subset \mathscr{F}\text{-closure of } \mathbf{I}_m \subset \mathbf{F}\text{-closure of } \mathbf{N}_m = \mathbf{F}_m;$$

the first inclusion follows from Exercise 4.10, whereas the second follows because $\mathbf{I}_m \subset \mathbf{N}_m$ and $\mathbf{F} \le \mathscr{F}$. Finally, $\mathscr{R}_m \subset \mathbf{R}_m$, because if $T \in \mathscr{R}_m$, then $T \in \mathscr{F}_m \subset \mathbf{F}_m$ and $\mathbf{M}(T) < \infty$.

4.16. (a)

$$T(f\,dx + g\,dy) = \sum_{k=1}^{\infty} \int_0^{2^{-k}} g(k^{-1}, y)\,dy.$$

$$\partial T(h) = \sum_{k=1}^{\infty} h(k^{-1}, 2^{-k}) - \sum_{k=1}^{\infty} h(k^{-1}, 0).$$

$$T \in \mathscr{R}_1.$$

(b)

$$T(f\,dx + g\,dy) = \int_0^1 \int_0^1 f(x, y)\,dx\,dy.$$

$$\partial T(h) = \int_0^1 h(1, y)\,dy - \int_0^1 h(0, y)\,dy.$$

$$T \in \mathbf{N}_1.$$

(c)

$$T(f\,dx + g\,dy) = \int_0^1 g(x, 0)\,dx.$$

$$\partial T(h) = \int_0^1 \frac{\partial h}{\partial y}(x, 0)\,dy.$$

$$T \in \mathscr{E}_1.$$

(See example following Theorem 4.7.)

(d)

$$T(f\,dx + g\,dy) = f(a).$$

$$\partial T(h) = \frac{\partial h}{\partial x}(a).$$

$$T \in \mathscr{E}_1.$$

(See Theorem 4.7.)

(e)

$$T(f\,dx + g\,dy) = \int_{\text{unit disc}} f(x, y)\,dx\,dy.$$

$$\partial T(h) = \int_0^1 h(\sqrt{1 - y^2}, y)\,dy - \int_0^1 h(-\sqrt{1 - y^2}, y)\,dy.$$

$$T \in \mathbf{N}_1.$$

4.17.

$$\mathscr{H}^1(E) = 1 - \sum_{n=1}^{\infty} 2^{n-1} \cdot 4^{-n} = 1/2.$$

Let T be the sum of all oriented intervals removed in defining E. Clearly, $T \in \mathscr{R}_1 - \mathbf{I}_1$. Therefore, $\mathscr{H}^1 \llcorner E \wedge \mathbf{i} = [0, 1] - T \in \mathscr{R}_1 - \mathbf{I}_1$.

4.18.

$$\partial T = (\partial T) \llcorner \{u > r\} + (\partial T) \llcorner \{u \leq r\}$$
$$= \partial[T \llcorner \{u > r\} + T \llcorner \{u \leq r\}].$$

4.19. Immediate from definition of $\langle T, u, r+ \rangle$.

4.20.

$$\mathbf{M}\langle T, u, r+ \rangle < \infty$$

for almost all r by (4).

$$\mathbf{M}(\partial \langle T, u, r+ \rangle) = \mathbf{M}\langle \partial T, u, r+ \rangle$$

(by (2))

$$< \infty$$

for almost all r by (4).

4.21. Choose currents A and B such that $T = A + \partial B$ and $\mathbf{F}(T) = \mathbf{M}(A) + \mathbf{M}(B)$. Since $\partial A = \partial T$, $A \in \mathbf{N}$. Since $\partial B = T - A$, $B \in \mathbf{N}$. Now

$$T \llcorner \{u \leq r\} = A \llcorner \{u \leq r\} + \partial[B \llcorner \{u \leq r\}] - \langle B, u, r+ \rangle$$

by 4.11(1). Therefore,

$$\mathbf{F}(T \llcorner \{u \leq r\}) \leq \mathbf{M}(A) + \mathbf{M}(B) + \mathbf{M}\langle B, u, r+ \rangle.$$

Integration and 4.11(4) yield (6) as desired.

4.22. First consider the case $\mathbf{M}(T) < \infty$. Given $\varepsilon > 0$, choose $\varphi \in \mathscr{D}^m$ with $||\varphi(x)||^* \leq 1$ such that $\mathbf{M}(T) \leq T(\varphi) + \varepsilon$. Then

$$\mathbf{M}(T) \leq T(\varphi) + \varepsilon = \lim T_i(\varphi) + \varepsilon$$
$$\leq \liminf \mathbf{M}(T_i) + \varepsilon.$$

Second, if $\mathbf{M}(T) = \infty$, given $\varepsilon > 0$, choose $\varphi \in \mathscr{D}^m$ with $||\varphi(x)||^* \leq 1$ such that $T(\varphi) > 1/\varepsilon$. Then

$$\liminf \mathbf{M}(T_i) \geq \lim T_i(\varphi) > 1/\varepsilon.$$

Hence, $\liminf \mathbf{M}(T_i) = \infty$, as desired.

4.23. We prove b, of which a is a special case.

$$f_\# S(\varphi) = S(f^\# \varphi) = \int_E \langle \vec{S}, f^\# \varphi \rangle l d\, \mathcal{H}^m$$

$$= \int_E \langle \wedge_m (Df(x))(\vec{S}), \varphi(f(x)) \rangle l(x) d\, \mathcal{H}^m x$$

$$= \int_E \left\langle \frac{\wedge_m (Df(x))(\vec{S})}{|\wedge_m (Df(x))(\vec{S})|}, \varphi(f(x)) \right\rangle l(x) \mathrm{ap} J_m(f\,|\,E) d\, \mathcal{H}^m x$$

(where the contribution from points at which $|\wedge_m (Df(x))(\vec{S})| = \mathrm{ap}\, J_m(f\,|\,E) = 0$ is still interpreted to be 0)

$$= \int_{f(E)} \sum_{y=f(x)} \left\langle \frac{(\wedge_m Df(x))(\vec{S})}{|(\wedge_m Df(x))(\vec{S})|}, \varphi \right\rangle l(x) d\, \mathcal{H}^m y$$

by the coarea formula, 3.13. Therefore,

$$f_\# S = (\mathcal{H}^m \llcorner f(E)) \wedge \sum_{y=f(x)} l(x) \frac{(\wedge_m Df(x))(\vec{S})}{|(\wedge_m Df(x))(\vec{S})|}$$

Chapter 5

5.1. Applying a homothety $\mu_r(x) = rx$ multiplies $\mathbf{M}(S)$ by r^{m+1}, $\mathbf{M}(T)$ by r^m, and hence both sides of the inequality by the same factor r^m.

5.2. Immediately from the definitions, $\mathbf{I}_{m+1} \subset \{T \in \mathcal{R}_{m+1} : \mathbf{M}(\partial T) < \infty\}$. The opposite inclusion follows from 5.4(1) because $\mathbf{I}_{m+1}$ is $\mathscr{F}$ dense in $\mathscr{F}_{m+1}$ (Exercise 4.10). Also from the definitions, $\mathcal{R}_m \subset \{T \in \mathscr{F}_m : \mathbf{M}(T) < \infty\}$. Conversely, suppose $T \in \mathscr{F}_m$ with $\mathbf{M}(T) < \infty$. Then $T = R + \partial S$, with $R \in \mathcal{R}_m$, $S \in \mathcal{R}_{m+1}$. Since $\mathbf{M}(\partial S) = \mathbf{M}(T - R) < \infty$, it follows from (2) that $S \in I_{m+1}$. Therefore, $T = R + \partial S \in \mathcal{R}_m$, as desired.

5.3. One good candidate is the sequence

$$T_k = \sum_{j=1}^{2^k} \left[\left(j - \frac{1}{2} \right) 2^{-k}, j2^{-k} \right] \in \mathbf{I}_1 \mathbf{R}^1,$$

which at first glance appears to converge to $\frac{1}{2}[0, 1] \notin \mathbf{I}_1 \mathbf{R}^1$.

$$\bullet \to \; \to \; \to \; \to \; \to \; \to \; \to \; \to \bullet$$
$$0 \qquad\qquad\qquad T_3 \qquad\qquad\qquad 1$$

5.4. Let T be the unit circle in $\mathbf{R}^2$ with multiplicity $1/N$. The only normal current S with $S = \partial T$ is the unit disc with multiplicity $1/N$. Since $\mathbf{M}(S) = \pi/N$ and $\mathbf{M}(T) = 2\pi/N$, the isoperimetric inequality does not hold for any constant γ.

Chapter 6

6.1. Just plug $f(y, z) = y \tan z$ into the minimal surface equation, 6.1.

6.2. Just plug $f(x, y) = \ln(\cos x / \cos y)$ into the minimal surface equation, 6.1.

6.3. Apply the minimal surface equation to $z = f(x, y)$. Let $w = u + iv$. Then

$$f_x = \frac{\partial f/\partial u}{\partial x/\partial u} + \frac{\partial f/\partial v}{\partial x/\partial v},$$

etc.

6.4. A surface of revolution has an equation of the form $r = g(z)$, where $r = \sqrt{x^2 + y^2}$. Differentiating $g^2 = x^2 + y^2$ implicitly yields

$$gg'z_x = x, \quad gg'z_y = y,$$

$$(g'^2 + gg'')z_x^2 + gg'z_{xx} = 1,$$

$$(g'^2 + gg'')z_y^2 + gg'z_{yy} = 1,$$

$$(g'^2 + gg'')z_x z_y + gg'z_{xy} = 0.$$

Applying the minimal surface equation to $z(x, y)$ yields

$$0 = [(1 + z_y^2)z_{xx} + (1 + z_x^2)z_{yy} - 2z_x z_y z_{xy}]gg'$$

$$= (1 + z_y^2)(1 - (g'^2 + gg'')z_x^2)$$

$$\quad + (1 + z_x^2)(1 - (g'^2 + gg'')z_y^2)$$

$$\quad + 2z_x z_y (g'^2 + gg'')z_x z_y$$

$$= 2 + (z_x^2 + z_y^2)(1 - g'^2 - gg'')$$

$$= (z_x^2 + z_y^2)(1 + g'^2 - gg'').$$

Therefore, $gg'' = 1 + g'^2$. Substituting $p = g'$ yields $gp(dp/dg) = 1 + p^2$. Integration yields $p^2 = ag^2 - 1$; that is,

$$\frac{dg}{\sqrt{a^2 g^2 - 1}} = \pm dz.$$

Integration yields $(1/a)\cosh^{-1} ag = \pm z + c$; that is, $r = g(z) = (1/a) \times \cosh(\pm az + c) = (1/a)\cosh(az \mp c)$, which is congruent to $r = (1/a)\cosh az$.

6.5.

$$0 = \operatorname{div} \frac{\nabla f}{1 + |\nabla f|^2}$$

$$= \frac{\partial}{\partial x}[f_x(1 + f_x^2 + f_y^2)^{-1/2}] + \frac{\partial}{\partial y}[f_y(1 + f_x^2 + f_y^2)^{-1/2}]$$

$$= f_{xx}(1 + f_x^2 + f_y^2)^{-1/2} - f_x(1 + f_x^2 + f_y^2)^{-3/2}(f_x f_{xx} + f_y f_{xy})$$

$$\quad + f_{yy}(1 + f_x^2 + f_y^2)^{-1/2} - f_y(1 + f_x^2 + f_y^2)^{-3/2}(f_x f_{xy} + f_y f_{yy}).$$

$$0 = (f_{xx} + f_{yy})(1 + f_x^2 + f_y^2) - f_x^2 f_{xx} - 2f_x f_y f_{xy} - f_y^2 f_{yy}$$

$$= (1 + f_y^2)f_{xx} - 2f_x f_y f_{xy} + (1 + f_x^2)f_{yy}.$$

6.6.

$$f(x, y) = (x^2 - y^2, 2xy).$$

$$(1 + |f_y|^2)f_{xx} - 2(f_x \cdot f_y)f_{xy} + (1 + |f_x|^2)f_{yy}$$

$$= (2 + 8|z|^2, 0) - 0$$

$$\quad + (-2 - 8|z|^2, 0)$$

$$= 0.$$

6.7. $g(z) = (u(x, y), v(x, y))$, satisfying the Cauchy–Riemann equations $u_x = v_y, u_y = -v_x$, and hence $u_{yy} = -u_{xx}$ and $v_{yy} = -v_{xx}$.

$$(1 + |f_y|^2)f_{xx} - 2(f_x \cdot f_y)f_{xy} + (1 + |f_x|^2)f_{yy}$$

$$= (1 + u_y^2 + v_y^2)(u_{xx}, v_{xx}) - 2(u_x u_y + v_x v_y)(u_{xy}, v_{xy})$$

$$\quad + (1 + u_x^2 + v_x^2)(u_{yy}, v_{yy})$$

$$= (1 + u_x^2 + u_y^2)(u_{xx}, v_{xx}) - 0 + (1 + u_x^2 + u_y^2)(-u_{xx}, -v_{xx})$$

$$= 0.$$

Chapter 8

8.1. To contradict 8.1, one might try putting a half-twist in the tail of Figure 8.1(1) so that the surface, like 8.1(2), would not be orientable. However, there is another surface like 8.1(2), in which the little hole goes in underneath in front and comes out on top in back. To contradict 8.4, one might try a large horizontal disc centered at the origin and a little vertical disc tangent to it at the origin:

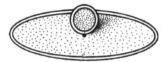

However, the area-minimizing surface is

8.2.

Front view

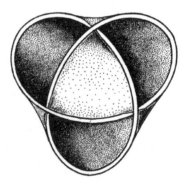

Top view

(There are two sheets in the middle.)

8.3. Let E be a unit element of surface area in $\mathbf{R}^4$ in a plane with orthonormal basis u, v. Let P, Q denote projection onto the x_1–x_2 and x_3–x_4 planes.

Then

$$\text{area } PE + \text{area } QE = |Pu \wedge Pv| + |Qu \wedge Qv|$$
$$\leq (|Pu||Pv| + |Qu||Qv|)$$
$$\leq (|Pu|^2 + |Qu|^2)^{1/2}(|Pv|^2 + |Qv|^2)^{1/2}$$
$$= |u||v| = 1 = \text{area } E.$$

Therefore, the area of any surface is at least the sum of the areas of its projections.

Now let S be a surface with the same boundary as the two discs $D_1 + D_2$. Since $\partial(P_\# S) = \partial D_1$ and $P_\# S$ and D_1 both lie in the x_1–x_2 plane, $P_\# S = D_1$. Similarly, $Q_\# S = D_2$. Therefore,

$$\text{area } S \geq \text{area } P_\# S + \text{area } Q_\# S$$
$$= \text{area } D_1 + \text{area } D_2$$
$$= \text{area } (D_1 + D_2).$$

Therefore, $D_1 + D_2$ is area minimizing.

Chapter 9

9.1. ∂R, where R is the pictured rectifiable current of infinitely many components C_j of length 2^{-j}.

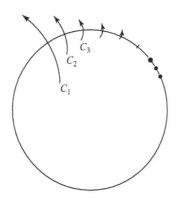

9.2. Suppose S is not area minimizing. For some $a, r > 0$ there is a rectifiable current T such that $\partial T = \partial(S \llcorner \mathbf{B}(a, r))$ and $\varepsilon = \mathbf{M}(S \llcorner \mathbf{B}(a, r)) - \mathbf{M}(T) > 0$. Choose j such that

$$\mathrm{spt}(S_j - S - (A + \partial B)) \cap \mathbf{U}(a, r + 1) = \varnothing,$$

$A \in \mathscr{R}_m$, $B \in \mathscr{R}_{m+1}$, and $\mathbf{M}(A) + \mathbf{M}(B) < \varepsilon$. Let $u(x) = |x - a|$, and apply slicing theory, 4.11(4), to choose $r < s < r + 1$ such that $\mathbf{M}\langle B, u, s+\rangle \leq \mathbf{M}(B)$. Now $S_j \llcorner \mathbf{B}(a, s)$ has the same boundary as

$$T + S \llcorner (\mathbf{B}(a, s) - \mathbf{B}(a, r)) + A \llcorner \mathbf{B}(a, s) - \langle B, u, s+\rangle$$

and more mass. Therefore, S_j is not area minimizing.

9.3. Suppose $p \in (\mathrm{spt}\, S) \cap \{\sqrt{x^2 + y^2} < R - 2\sqrt{R}\}$. Then the distance from p to spt ∂S exceeds $2\sqrt{R}$. By monotonicity,

$$\mathbf{M}(S) > \pi(2\sqrt{R})^2 = 4\pi R = \text{area cylinder}.$$

9.4. False:

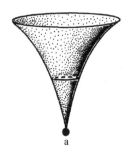

a

9.5. (a) One example is two unit orthogonal (complex) discs in $\mathbf{R}^4$, where the density jumps up to 2 at the origin (cf. 6.3 or Exercise 8.3).

(b) Suppose $x_i \rightarrow x$ but $f(x) < \overline{\lim} f(x_i)$. Choose $0 < r_0 < \mathrm{dist}(x, \mathrm{spt}\, \partial T)$ such that $\Theta^m(T, x, r_0) < \overline{\lim} f(x_i)$.

Choose $0 < r_1 < r_0$ such that

$$r_0^m \Theta^m(T, x, r_0) < r_1^m \overline{\lim} f(x_i).$$

Choose i such that $|x_i - x| < r_0 - r_1$ and

$$r_0^m \Theta^m(T, x, r_0) < r_1^m \Theta^m(T, x_i).$$

By monotonicity,

$$r_0^m \Theta^m(T, x, r_0) < r_1^m \Theta^m(T, x_i, r_1).$$

However, since $\mathbf{B}(x, r_0) \supset \mathbf{B}(x_i, r_1)$,

$$r_0^m \Theta^m(T, x, r_0) \geq r_1^m \Theta^m(T, x_i, r_1).$$

This contradiction proves that f is upper semicontinuous.

9.6. Suppose $x \in \mathbf{S}(0, 1) - \mathrm{Tan}(\mathrm{spt}\, T, \mathbf{0})$. Then for some $\varepsilon > 0$, for all sufficiently large r,

$$\mathrm{spt}\, \mu_{r\#}T \cap \mathbf{B}(x, \varepsilon) = \varnothing.$$

Consequently, $x \notin \mathrm{spt}\, C$.

Figure 9.7.2 shows an example in which $\mathrm{spt}\, C \neq \mathrm{Tan}(\mathrm{spt}\, T, \mathbf{0})$.

9.7. $$C - D = \lim \mu_{r_j/s_j\#}[C - \mu_{r_j^{-1}\#}T] = 0.$$

9.8. (a) Let T_n be the homothetic expansion of T by 2^{n^2-n} (n odd). T_n consists of the interval $[2^{-n}, 2^{n-1}]$ on the x-axis, stuff outside $\mathbf{B}(0, 2^{n-2})$, and stuff inside $\mathbf{B}(0, 2^{-n})$ of total mass less than 2^{-n+2}. As a limit of the T_n, the nonnegative x-axis is an oriented tangent cone. Similarly, taking n even yields the y-axis.

(b) Let S_n be the homothetic expansion of T by 2^{n^2} (n odd). S_n consists of the interval $[1, 2^{2n-1}]$ on the x-axis, the segment from (0, 1) to (1,0), the interval $[2^{-2n-1}, 1]$ on the y-axis, stuff outside $\mathbf{B}(0, 2^{2n-2})$, and stuff inside $\mathbf{B}(0, 2^{-2n-1})$ of total mass less than 2^{-2n+1}. S_n converges to the interval $[1, \infty)$ on the x-axis, plus the segment from (0, 1) to (1,0), plus the interval (0, 1) on the y-axis. This limit is not a cone.

(c) Looking at balls of radius 2^{-n^2} (n odd) exhibits the lower density 0. Looking at balls of radius 2^{-n^2} (n even) exhibits the upper density $1/2$. (Of course, for a subset of the nonnegative x-axis, the densities must lie between 0 and $1/2$.)

Chapter 10

10.1.

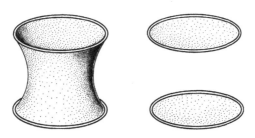

or see Figure 6.1.4.

10.2. False. $[(-1, 0, 0), (1, 0, 0)] + [(0, -1, 100), (0, 1, 100)] \in$ $\mathscr{R}_1 \mathbf{R}^2 \times \mathbf{R}^1$ is area minimizing, but its projection, $[(-1, 0), (1, 0)] + [(0, -1), (0, 1)]$, is not.

10.3.

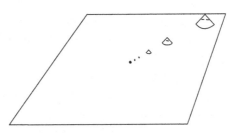

Chapter 11

11.1. Four similarly oriented, parallel circles

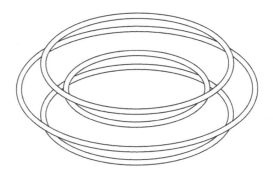

bound a catenoid and a horizontal annulus with cross-sections:

11.2. Because the tangent plane is constant, $\mathbf{v}(S)(\mathbf{R}^n \times G_2\mathbf{R}^n) = $ area $S = \pi$.

11.3. Two close parallel unit discs, which can be deformed to a surface like that in Figure 11.1.3. Any unstable minimal surface, as in Figure 6.1.2, which can be deformed to a surface in Figure 6.1.4.

11.4. The surface of Figure 11.1.3. The catenoid has less area.

Chapter 13

13.1. Consider two disjoint discs of area 1/2 and alter the metric to make their perimeters very small.

13.2. There is no least-perimeter enclosure of given area a because for large n, area a/n on n cusp hills can be enclosed with total perimeter $n(a/n)^2 = a/n$. Thus, perimeter 0 can be approached but not realized.

13.3. The basic computation is that if two lines meet at an angle of less than 120 degrees, they can profitably merge for a little while. It follows that lines meet at most in threes. Of course, they cannot meet in twos (except at 180 degrees, when they are really just one line) except at given boundary points, where the angle must still be at least 120 degrees.

Chapter 14

14.1. Use a decomposition such as Figure 14.8.1, but with four contributions: each of the three single components and the whole.

14.2. (a) If the exterior had a bounded component, you could knock out the interface with another region, decreasing perimeter and maintaining areas *at least* A_1, A_2.

(b) Bumping would contradict regularity. If it bumps nothing, it is two circular arcs on a circle—the standard double bubble. (If instead of sliding you use reflection, you do not need to assume that the outermost component lies on a circular arc.)

(c) Among standard double bubbles with areas at least A_1, A_2, the one with areas exactly A_1, A_2 has least perimeter and therefore solves the prescribed area problem.

Chapter 15

15.1. By the two facts, the perimeter P satisfies

$$P \geq (n/2)(2\sqrt{\pi}) + (1/2)(2\sqrt{n\pi}).$$

The perimeter-to-area ratio for the regular hexagonal tiling is easily computed as $12^{1/4}$. The problem is finished by verifying the algebraic inequality that for $n \leq 398$,

$$P/n \geq (1 + 1\sqrt{n})\sqrt{\pi} \geq 12^{1/4}.$$

Chapter 18

18.1. By 18.3(1), for the outward unit normal, $H_\psi = H - d\psi/dn = -1 - -1 = 0$.

18.2. The left-hand side is $(4/3)\pi((1+r)^3 - 1)$. On the right-hand side, $H = -1$, $s_\delta(t) = t$, $c_\delta(t) = 1$, the integrand is $(1+t)^2$, the inside integral is $((1+r)^3 - 1)/3$, and the outside integral is $(4/3)\pi((1+r)^3 - 1)$.

18.3. As in 18.2, the left-hand side is $(4/3)\pi((1+r)^3 - 1)$. On the right-hand side, $H = -1$, the integrand is $e^{2t} > 1 + 2t + 2t^2 > (1+t)^2$, so that, as in 18.2, the outside integral is greater than $(4/3)\pi((1+r)^3 - 1)$.

18.4. The sphere has radius $1/2$ and density $1/\pi$. The cost of enclosing half the volume is the weighted length of an equator, $(1/\pi)(2\pi/2) = 1$. The model sphere has radius 1 and density $1/4\pi$. Its cost for enclosing half the volume is $(1/4\pi)(2\pi) = 1/2$.

18.5. The unit sphere has Ricci curvature 2 and density $1/4\pi$. The cost of enclosing half the volume is the weighted length of an equator, $(1/4\pi)(2\pi) = 1/2$. The cost in the model Gauss line is just the density at the origin, namely $1/\sqrt{2\pi}$ or approximately .398.

18.6. The area of a sphere satisfies $A(r) = 4\pi r^2$, $A'(r) = 8\pi r$, $A''(r) = 8\pi$. To recover these results from the variational formulae, first note that $H = -1/r$ and $II^2 = 2/r^2$. In the first variational formula, the integrand is $2/r$, yielding $8\pi r$. In the second variational formula, the integrand is $0 + 4/r^2 - 2/r^2 - 0 - 0 = 2/r^2$, yielding 8π.

18.7. The area of $\{x = t\}$ satisfies $A(t) = (2\pi)^{-.5} \exp(-.5t^2)$, $A'(t) = -tA(t)$, $A''(a) = (t^2 - 1) A(t)$. To recover these results from the variational formulae, first note $H_\psi = 0 + t/2$ and $II^2 = 0$. In the first variational formula, the integrand is t, yielding $-tA(t)$. In the second variational formula, the integrand is $0 + t^2 - 0 - 0 + 1$, yielding $(t^2 - 1)A(t)$.

18.8. In $\mathbf{R}^3$, $\gamma = 0$ and spheres minimize perimeter, so $P(V) = c(V)^{2/3}$, $P' = (2/3)cV^{-1/3}$, $P'' = -(2/9)cV^{-4/3}$. Inequality 18.10(1) states that $-(2/9)c^2V^{-2/3} \leq 0$, which holds strictly. In G^3, $\gamma = -1$ and planes $\{x = t\}$ minimize perimeter. So $P = c \exp(-t^2/2)$, $dP/dt = -tP$, $d^2P/dt^2 = (t^2 - 1)P$. Since $dV/dt = P$, by the chain rule $dP/dV = -t$ and $d^2P/dV^2 = -1/P$, yielding equality in 18.10(1).

18.9. Using the data from the solution to 18.8, inequality 18.10(2) states that $-(2/9)c^2V^{-2/3} \leq -(4/9)c^2V^{-2.3}/2 - 0$, which holds with equality.

Chapter 19

19.1. They meet where the circle and the band have the same perimeter. This occurs when $\sqrt{4\pi A} = 2$; that is, when A is $1/\pi$ (or $1 - 1/\pi$).

19.2. See [Corneli, Holt, *et al.*, Prop. 5.1.1].

Chapter 20

20.1. For horizontal translation, the normal component of velocity is given by $u = \cos\theta$, and $|\nabla u|^2 = \sin^2\theta$. Therefore

$$P'' = \int |\nabla u|^2 - \mathrm{II}^2 u^2 + u^2(\partial^2\psi/\partial n^2)$$

$$= \int \sin^2\theta - \cos^2\theta + u^2(\partial^2\psi/\partial n^2) = 0 + \int u^2(\partial^2\psi/\partial n^2) \geq 0.$$

20.2. This density is not smooth at the origin.

20.3. If the density is radial, then the generalized mean curvature is constant, so the first variation vanishes. On the other hand, if the mean curvature is constant, then along every ray $\partial\psi/\partial r$ is constant on every sphere about the origin, so ψ is constant on every sphere about the origin and the density is radial.

20.4. Since the density is a product, its $\log\psi$ as a function of r^2 is of the form

$$\psi(r^2) = f(x) + g(y).$$

Differentiation with respect to x yields

$$\psi'(r^2)(2x) = f'(x),$$

and similarly

$$\psi'(r^2)(2y) = f'(y).$$

Therefore $f'(x)/2x$ and $f'(y)/2y$ are equal and constant. Integration yields the result.

20.5. A sphere smoothly flattened about the origin.

20.6. We may assume that the circle is given in polar coordinates by $r = 2a\cos\theta$. The outward normal is at angle θ with $\nabla\psi$, so $d\psi/dn$ is $|\nabla\psi|\cos\theta = (p/r)\cos\theta = p/2a$, a constant. Since κ is constant too, the generalized curvature is constant.

20.7. By symmetry the first variation vanishes. We consider translation in the x_n direction, which by symmetry preserves volume to first order. In terms of the polar angle φ, as in the solution to Exercise 20.1, the normal component of velocity is given by $u = \cos \varphi$, and $|\nabla u|^2 = \sin^2 \varphi$. Therefore the first two terms for P'' net 0, as they must since with unit density neither volume nor perimeter varies under translation. The last term is negative because with $\psi = p \log r$, $\partial^2 \psi / \partial n^2 = \psi'' = -p/r^2 < 0$.

20.8. For given perimeter, a ball has maximum unweighted volume, and a ball about the origin has maximum weighted volume.

Bibliography

Section numbers follow the entries, indicating the locations in this book where the publications are cited.

Allard, William K. On the first variation of a varifold. *Ann. Math.* **95** (1972), 417–491. (§§5.3, 9.3, 10.2, 11.2, 17.2)

Almgren, F. J., Jr.

[1] Existence and regularity almost everywhere of solutions to elliptic variational problems with constraints. *Mem. AMS* **165** (1976). (§§8.6, 11.3, 13.3, 13.4, 13.8, 13.10)

[2] Optimal isoperimetric inequalities. *Indiana Univ. Math. J.* **35** (1986), 451–547. (§§5.3, 17.1)

[3] Q-valued functions minimizing Dirichlet's integral and the regularity of area minimizing rectifiable currents up to codimension two. *Bull. AMS* **8** (1983), 327–328. (§8.3)

[4] Questions and answers on area minimizing surfaces and geometric measure theory. *Proc. Symp. Pure Math.* **54** (1993), Part 1, 29–53. (Preface)

[5] Review of "Geometric Measure Theory: A Beginner's Guide." *Am. Math. Monthly* **96** (1989), 753–756. (Preface)

[6] Some interior regularity theorems for minimal surfaces and an extension of Bernstein's theorem. *Ann. Math.* **84** (1966), 277–292. (§8.1)

Almgren, F. J., Jr., and **Taylor**, J. E. Geometry of soap films. *Sci. Am.* **235** (1976), 82–93. (§§13.8, 13.9)

Almgren, Fred, **Taylor**, Jean E., and **Wang**, Lihe. Curvature-driven flows: a variational approach. *Siam J. Control Optimization* **31** (1993), 387–437. (§16.7)

Ambrosio, Luigi, and **Braides**, Andrea. Functionals defined on partitions in sets of finite perimeter II: semicontinuity, relaxation and homogenization. *J. Math. Pures Appl.* **69** (1990), 307–333. (§16.4)

Ambrosio, Luigi, **Fusco**, Nicola, and **Pallara,** Diego. "Functions of Bounded Variation and Free Discontinuity Problems." Oxford University Press, Oxford, 2000. (§12.4)

Ambrosio, Luigi, and **Kirchheim**, Bernd. Rectifiable sets in metric and Banach spaces. *Math. Ann.* **318** (2000), 527–555. (§§1.8, 4.12)

Anderson, Michael T. Geometrization of 3-manifolds via the Ricci flow. *Notices AMS* **51** (2004), 184–193. (§18.11)

Bakry, D., and **Émery**, M. Diffusions hypercontractives. *Lecture Notes Math.* **1123** (1985), 177–206. (§§18.0, 18.3)

Bakry, D., and **Ledoux**, M. Lévy–Gromov's isoperimetric inequality for an infinite dimensional diffusion generator. *Invent. Math.* **123** (1996), 259–281. (§18.3)

Bates, S. M. Toward a precise smoothness hypothesis in Sard's theorem. *Proc. AMS* **117** (1993), 279–283. (§12.1)

Bayle, Vincent. Propriétés de concavité du profil isopérimétrique et applications. Graduate thesis, Institut Fourier, Universite Joseph-Fourier, Grenoble I, 2004. (Chapter 18)

Berger, M.

[1] "Geometry," Vol. II. Springer-Verlag, New York, 1972. (§13.2)

[2] Quelques problèmes de géométrie riemannienne ou deux variations sur les espaces symétriques compacts de rang un. *Enseignement Math.* **16** (1970), 73–96. (§6.5)

[3] "A Panoramic View of Riemannian Geometry." Springer-Verlag, Berlin, 2003. (§13.2)

Besicovitch, A. S. A general form of the covering principle and relative differentiation of additive functions I, II. *Proc. Cambridge Phil. Soc.* **41** (1945), 103–110; **42** (1946), 1–10. (§2.7)

Betta, M. Francesca, **Brock**, Friedemann, **Mercaldo**, Anna, and **Posteraro**, M. Rosaria. Weighted isoperimetric inequalities on R^n and applications to rearrangements. *Math. Nachr.* **281** (2008), 466–498. (§20.5)

Bieberbach, Ludwig. Über eine Extremaleigenschaft des Kreises. *J.-ber. Deutsch. Math.-Verein.* **24** (1915), 247–250. (§§2.8, 14.3)

Blaschke, Wilhelm. "Kreis und Kugel." Veit & Comp., Leipzig, 1916. (§2.8)

Bombieri, E., **De Giorgi**, E., and **Giusti**, E. Minimal cones and the Bernstein problem. *Invent. Math.* **7** (1969), 243–268. (§§8.1, 10.7)

Borell, Christer.

[1] The Brunn–Minkowski inequality in Gauss space. *Invent. Math.* **30** (1975), 207–216. (§18.2)

[2] Geometric inequalities in option pricing. "Convex Geometric Analysis" (K. M. Ball and V. Milman, Eds.). Cambridge University Press, Cambridge, UK, 1999, Vol. 5, 29–51. (§18.2)

Boyer, Wyatt, **Chambers**, Gregory R., **Loving**, Alyssa, and **Tammen**, Sarah. Isoperimetric regions in R^n with density r^p. arXiv.org (2015). (§20.7)

Boys, C. V. "Soap-Bubbles: Their Colors and the Forces Which Mold Them." Dover, New York, 1959. (§14.0)

Brakke, Kenneth A.

[1] Minimal cones on hypercubes. *J. Geom. Anal.* **1** (1991), 329–338. (§6.5)

[2] Polyhedral minimal cones in R^4. Preprint (1993). (§§6.5, 13.10)

[3] Soap films and covering spaces. *J. Geom. Anal.* **5** (1995), 445–514. (§6.5)

[4] The surface evolver. *Exp. Math.* **1** (1992), 141–165. (§§13.0, 13.12, 13.13, 16.7)

[5] Century-old soap bubble problem solved! *Imagine That!* **3** (Fall, 1993). The Geometry Center, University of Minnesota, 1–3. (§13.13)

Bray, Hubert. Proof of the Riemannian Penrose conjecture using the positive mass theorem. *J. Diff. Geom.* **59** (2001), 177–267. (Preface, §16.7)

Bray, Hubert, and **Morgan**, Frank. An isoperimetric comparison theorem for Schwarzschild space and other manifolds. *Proc. AMS* **130** (2002), 1467–1472. (§13.2)

Brothers, J. E. (Ed.). Some open problems in geometric measure theory and its applications suggested by participants of the 1984 AMS Summer Institute. "Geometric Measure Theory and the Calculus of Variations" (W. K. Allard and F. J. Almgren, Jr., Eds.). *Proc. Symp. Pure Math.* **44** (1986), 441–464. (Preface)

Brothers, John E., and **Morgan**, Frank. The isoperimetric theorem for general integrands. *Michigan Math. J.* **41** (1994), 419–431. (§13.2)

Brubaker, Nicholas D., **Carter**, Stephen, **Evans**, Sean M., **Kravatz**, Daniel E., Jr., **Linn**, Sherry, **Peurifoy**, Stephen W., and **Walker**, Ryan. Double bubble experiments in the three-torus. *Math Horizons* (May, 2008), 18–21. (§19.7)

Burago, Yu. D., and **Zalgaller**, V. A. "Geometric Inequalities." Springer-Verlag, New York, 1988. (§§2.8, 13.2, 18.4, 18.6)

Capogna, Luca, **Danielli**, Donatella, **Pauls**, Scott D., and **Tyson**, Jeremy T. "An Introduction to the Heisenberg Group and the Sub-Riemannian Isoperimetric Problem." Birkhäuser, Boston, 2007. (§13.2)

Carlen, E. A., and **Kerce**, C. On the cases of equality in Bobkov's inequality and Gaussian rearrangement. *Calc. Var. PDE* **13** (2001), 1–18. (§18.2)

Caroccia, M., and **Maggi**, F. A sharp quantitative version of Hales' isoperimetric honeycomb theorem. arXiv.org (2014). (§15.1)

Carrión Álvarez, Miguel, **Corneli**, Joseph, **Walsh**, Genevieve, and **Behesht**, Shabnam. Double bubbles in the three-torus. *Exp. Math.* **12** (2003), 79–89. (§§14.24, 19.7)

Carroll, Colin, **Jacob**, Adam, **Quinn**, Conor, and **Walters**, Robin. The isoperimetric problem on planes with density. *Bull. Austral. Math. Soc.* **78** (2008), 177–197. (§18.3, Exercise 18.1)

Cashman, K. V., **Rust**, A. C., and **Wright**, H. M. Pattern and structure of basaltic reticulite: foam formation in lava fountains. *Geophys. Res. Abstr.* **8** (2006), 05398. (§15.12)

Chambers, Gregory R. Proof of the Log-Convex Density Conjecture. arXiv.org (2013). (§20.6)

Chang, Sheldon. Two dimensional area minimizing integral currents are classical minimal surfaces. *J. AMS* **1** (1988), 699–778. (§8.3)

Chang, Sun-Yung A., **Gursky**, Matthew J., and **Yang**, Paul. Conformal invariants associated to a measure. *Proc. Natl. Acad. Sci. USA* **103** (2006), 2535–2540. (§18.3)

Chavel, Isaac.

[1] "Isoperimetric Inequalities." Cambridge University Press, Cambridge, UK, 2001. (§5.3)

[2] "Riemannian Geometry: A Modern Introduction." Cambridge University Press, Cambridge, UK, 1993. (§18.8)

Cianchi, A., **Fusco**, N., **Maggi**, F., and **Pratelli**, A. On the isoperimetric deficit in Gauss space. *Amer. J. Math.* **133** (2011), 131–186. (§18.2)

Cipra, Barry. Why double bubbles form the way they do. *Science* **287** (17 March 2000), 1910–1911. (§14.0)

Coifman, R. R., and **Weiss**, G. Extensions of Hardy spaces and their use in analysis. *Bull. AMS* **83** (1977), 569–645. (§18.1)

Conway, John H., and **Sloane**, Neil. The cell structures of certain lattices. "Miscellanea Mathematica" (Peter Hilton, Friedrich Hirzebruch, and Reinhold Remmert, Eds.). Springer-Verlag, New York, 1993. (§15.11)

Corneli, Joseph, **Corwin**, Ivan, **Hoffman**, Neil, **Hurder**, Stephanie, **Sesum**, Vojislav, **Xu**, Ya, **Adams**, Elizabeth, **Davis**, Diana, **Lee**, Michelle, and **Visocchi**, Regina. Double bubbles in Gauss space and spheres. *Houston J. Math.* **34** (2008), 181–204. (§§19.4, 19.5)

Corneli, Joseph, **Hoffman**, Neil, **Holt**, Paul, **Lee**, George, **Leger**, Nicholas, **Moseley**, Stephen, and **Schoenfeld**, Eric. Double bubbles in S^3 and H^3. *J. Geom. Anal.* **17** (2007), 189–212. (§19.4)

Corneli, Joseph, **Holt**, Paul, **Lee**, George, **Leger**, Nicholas, **Schoenfeld**, Eric, and **Steinhurst**, Benjamin. The double bubble problem on the flat two-torus. *Trans. AMS* **356** (2004), 3769–3820. (§§14.24, 19.6)

Corwin, Ivan, **Hoffman**, Neil, **Hurder**, Stephanie, **Sesum**, Vojislav, and **Xu**, Ya. Differential geometry of manifolds with density. *Rose–Hulman Und. Math. J.* **7**(1) (2006). (§18.3)

Costa, C. "Imersões Minimas Completas em $\mathbf{R}^3$ de Gênero um e Curvatura Total Finita." Doctoral thesis. IMPA. Rio de Janeiro, Brasil, 1982; Example of a complete minimal immersion in $\mathbf{R}^3$ of genus one and three embedded ends. *Bull. Soc. Bras. Mat.* **15** (1984), 47–54. (Figure 6.1.3a)

Cotton, Andrew, and **Freeman**, David. The double bubble problem in spherical and hyperbolic space. *Int. J. Math. Math. Sci.* **32** (2002), 641–699. (§§14.24, 19.1–19.4)

Courant, R., and **Hilbert**, D. "Methods of Mathematical Physics," Vol. I, English ed. Wiley, New York, 1953. (§14.14)

Cox, Christopher, **Harrison**, Lisa, **Hutchings**, Michael, **Kim**, Susan, **Light**, Janette, **Mauer**, Andrew, and **Tilton**, Meg.

[1] The shortest enclosure of three connected areas in $\mathbf{R}^2$. *Real Anal. Exch.* **20** (1994/1995), 313–335. (§§13.1, 13.12, 14.23)

[2] The standard triple bubble type is the least-perimeter way to enclose three connected areas. NSF "SMALL" undergraduate research Geometry Group report. Williams College, Williamstown, MA, 1992. (§14.8)

Cox, S. J. Calculations of the minimal perimeter for N deformable cells of equal area confined in a circle. *Phil. Mag. Lett.* **86** (2006), 569–578. (§13.1)

Cox, S. J., and **Graner**, F. Large two-dimensional clusters of equal-area bubbles: the influence of the boundary in determining the minimum energy configuration. *Phil. Mag.* **83** (2003), 2573–2584. (§13.1)

Cox, S. J., **Graner**, F., **Vaz**, M. Fátima, **Monnereau-Pittet**, C., and **Pittet**, N. Minimal perimeter for N identical bubbles in two dimensions: calculations and simulations. *Phil. Mag.* **83** (2003), 1393–1406. (§13.1)

Coxeter, H. S. M. "Regular Polytopes." Dover, New York, 1973. (§15.11)

Croke, Christopher B. Some isoperimetric inequalities and eigenvalue estimates. *Ann. Scient. Éc. Norm. Sup.* **13** (1980), 419–435. (§17.3)

Daily, Marilyn. Proof of the double bubble curvature conjecture. *J. Geom. Anal.* **17** (2007), 75–85. (§14.14)

Dahlberg, Jonathan, **Dubbs**, Alexander, **Newkirk**, Edward, and **Tran**, Hung. Isoperimetric regions in the plane with density r^p. *New York J. Math.* **16** (2010), 31–51. (§20.7)

Dao Trong Thi. Minimal real currents on compact Riemannian manifolds. *Izv. Akad. Nauk. SSSR Ser. Mat.* **41** (1977) [English translation in *Math. USSR Izv.* **11** (1977), 807–820]. (§6.5)

David, Guy. Should we solve Plateau's problem again? "Advances in Analysis: The Legacy of Elias M. Stein" (Charles Fefferman, Alexandru D. Ionescu, DH Phong, and Stephen Wainger, Eds.), *Princeton Mathematical Series* **50** (2012), 108–145. (§13.10)

De Giorgi, E.

[1] "Frontiere Orientate di Misura Minima" (*Sem. Mat. Scuola Norm. Sup. Pisa*, 1960–1961). Editrice Tecnico Scientifica, Pisa, 1961. (§§1.0, 4.0)

[2] Sulla proprietà isoperimetrica dell'ipersfera, nella classe degli insiemi aventi frontiera orientata di misura finita. *Mem. Acc. Naz. Lincei*, Ser. 8, **5** (1958), 33–44. (§13.2)

[3] Su una teoria generale della misura $r - 1$ dimensionale in un spazio ad r dimensioni. *Ann. Mat.* **4** (1955), 95–113. (§§1.0, 4.0)

[4] "Ennio De Giorgi: Selected Papers" (L. Ambrosio, G. Dal Maso, M. Forti, M. Miranda, and S. Spagnolo, Eds.). Springer, Berlin, 2006. (§§1.0, 4.0)

De Lellis, Camillio. Rectifiable sets, densities, and tangent measures. "Zürich Lectures in Advanced Mathematics". European Mathematical Society (EMS), Zürich, 2008. (§3.12)

De Pauw, Thierry. Autour du théorème de la divergence. *Panor. Synthèses* **18** (2004), 85–121 (§12.2).

De Pauw, Thierry, and **Hardt**, Robert. Size minimization and approximating problems. *Calc. Var. PDE* **17** (2003), 405–442. ([Morgan 17])

De Philippis, Guido, **Franzina**, Giovanni, and **Pratelli**, Aldo. Existence of isoperimetric sets with densities converging from below on $\mathbf{R}^n$. arXiv.org (2014). (§20.5)

de Rham, Georges.

[1] "On the Area of Complex Manifolds. Notes for the Seminar on Several Complex Variables." Institute for Advanced Study, Princeton, NJ, 1957–1958. (§6.5)

[2] Variétés différentiables, formes, courants, formes harmoniques. *Act. Sct. Indust.* **1222** (1955). (§4.0)

Douglas, Jesse. Solution of the problem of Plateau. *Trans. AMS* **33** (1931), 263–321. (§1.2)

Falconer, K. J. "The Geometry of Fractal Sets." Cambridge University Press, Cambridge, UK, 1985. (§3.17)

Federer, Herbert.

[1] **"Geometric Measure Theory."** Springer-Verlag, New York, 1969. (cited throughout)

[2] The singular sets of area minimizing rectifiable currents with codimension one and of area minimizing flat chains modulo two with arbitrary codimension. *Bull. AMS* **76** (1970), 767–771. (§§8.2, 11.1)

[3] Some theorems on integral currents. *Trans. AMS* **117** (1965), 43–67. (§6.5)

Federer, Herbert, and **Fleming**, Wendell H. Normal and integral currents. *Ann. Math.* **72** (1960), 458–520. (§§1.0, 4.0)

Fejes Tóth, G. An isoperimetric problem for tessellations. *Studia Sci. Math. Hungarica* **10** (1975), 171–173. (§15.9)

Fejes Tóth, L.

[1] An arrangement of two-dimensional cells. *Ann. Univ. Sci. Budapest., Sect. Math.* **2** (1959), 61–64. (§15.1)

[2] "Lagerungen in der Ebene auf der Kugel und im Raum." Die Grundlehren der Math. Wiss., Vol. 65. Springer-Verlag, Berlin, 1953, 1972. (§15.1)

[3] "Regular Figures," International Series of Monographs on Pure and Applied Mathematics Vol. 48. Macmillan, New York, 1964. (§15.1)

[4] Über das kürzeste Kurvennetz das eine Kugeloberfläche in Flächengleiche konvexe Teile zerlegt. *Mat. Term. Ertesitö* **62** (1943), 349–354. (§15.1)

[5] What the bees know and what they do not know. *Bull. AMS* **70** (1964), 468–481. (§15.8)

Feynman, Richard P. The internal geometry of crystals. "The Feynman Lectures on Physics," Vol. II. Addison-Wesley, Reading, MA, 1963–1965, Chapter 30. (§16.4)

Figalli, Allesio, and **Maggi**, Francesco. On the isoperimetric problem for radial log-convex densities. *Calc. Var. Partial Differential Equations* **48** (2013), 447–489. (§20.5)

Fleming, Wendell H.

[1] Flat chains over a finite coefficient group. *Trans. AMS* **121** (1966), 160–186. (§16.2)

[2] On the oriented Plateau problem. *Rend. Circ. Mat. Palermo* (2)**11**, (1962), 1–22. (§8.1)

Foisy, Joel. "Soap Bubble Clusters in $\mathbf{R}^2$ and $\mathbf{R}^3$." Undergraduate thesis. Williams College, Williamstown, MA, 1991. (§§13.0, 14.0, 14.3, 14.9)

Foisy, Joel, **Alfaro**, Manuel, **Brock**, Jeffrey, **Hodges**, Nickelous, and **Zimba**, Jason. The standard double soap bubble in $\mathbf{R}^2$ uniquely minimizes perimeter. *Pac. J. Math.* **159** (1993), 47–59. Featured in the 1994 AMS "What's Happening in the Mathematical Sciences." (Preface, §§13.1, 14.0, 14.3, 14.10)

Fomenko, A. T. "The Plateau Problem. I: Historical Survey. II: Present State of the Theory." Gordon & Breach, New York, 1990. (Preface)

Fortes, M. A., and **Rosa**, M. Emília. The surface energy and contact angles of a liquid foam. *J. Colloid and Interface Sci.* **241** (2001), 205–214. (§13.1)

Fortes, M. A., and **Teixeira**, P. I. C. Minimum perimeter partitions of the plane into equal numbers of regions of two different areas. *Euro. Phys. J. E (Soft Matter)* **6** (2001), 133–137. (§13.1)

Francis, George, **Sullivan**, John M., **Kusner**, Rob B., **Brakke**, Ken A., **Hartman**, Chris, and **Chappell**, Glenn. The minimax sphere eversion. "Visualization and Mathematics." Springer, New York, 1997, 3–20. (§16.7)

Futer, David, **Gnepp**, Andrei, **McMath**, David, **Munson**, Brian A., **Ng**, Ting, **Pahk**, Sang-Hyoun, and **Yoder**, Cara. Cost-minimizing networks among immiscible fluids in $\mathbf{R}^2$. *Pac. J. Math.* **196** (2000), 395–414. (§16.3)

Giusti, Enrico. "Minimal Surfaces and Functions of Bounded Variation." Birkhäuser, Boston, 1984. (Preface, §§4.0, 12.4)

Gluck, Herman, **Mackenzie**, Dana, and **Morgan**, Frank. Volume-minimizing cycles in Grassmann manifolds. *Duke Math. J.* **79** (1995), 335–404. (§6.5)

Gonick, Larry, Fill'er up. *Discover Magazine*, August 1994, 80–81. (§15.10)

Graner, F., **Jiang**, Y., **Janiaud**, E., and **Flament**, C. Equilibrium states and ground state of two-dimensional fluid foam. *Phys. Rev. E* **63** (2001), 11402-1-13. (§13.12)

Grayson, Matthew A. Shortening embedded curves. *Ann. Math.* **129** (1989), 71–111. (§16.7)

Gromov, M.

[1] Isoperimetric inequalities in Riemannian manifolds. Appendix I to "Asymptotic Theory of Finite Dimensional Normed Spaces" by Vitali D. Milman and Gideon Schechtman. Lecture Notes in Mathematics, No. 1200, Springer-Verlag, New York, 1986. (§§13.2, 18.4)

[2] Isoperimetry of waists and concentration of maps. *Geom. Funct. Anal.* **13** (2003), 178–215. (§18.1)

Grünbaum, Branko, and **Shephard**, G. C. "Tilings and Patterns." Freeman, New York, 1987. (Figure 15.9.1)

Grüter, Michael.

[1] Free boundaries in geometric measure theory and applications. "Variational Methods for Free Surface Interfaces" (P. Concus and R. Finn, Eds.). Springer-Verlag, New York, 1986. (§12.3)

[2] Optimal regularity for codimension one minimal surfaces with a free boundary. *Manuscript Math.* **58** (1987), 295–343. (§12.3)

[3] Boundary regularity for solutions of a partitioning problem. *Arch. Rational Mech. Anal.* **97** (1987), 261–270. (§12.3)

Hales, Thomas C. The honeycomb conjecture. *Disc. Comp. Geom.* **25** (2001), 1–22. (Preface, Chapter 15)

Hardt, Robert, and **Simon**, Leon.

[1] Boundary regularity and embedded solutions for the oriented Plateau problem. *Ann. Math.* **110** (1979), 439–486. (§8.4)

[2] "Seminar on Geometric Measure Theory." Birkhäuser, Boston, 1986. (Preface).

Harrison, Jenny. Stokes' theorem for nonsmooth chains. *Bull. AMS* **29** (1993), 235–242. (§12.2)

Harvey, F. Reese. "Spinors and Calibrations." *Perspectives in Mathematics* **9**, Academic Press, Boston, 1990. (§6.5)

Harvey, Reese, and **Lawson**, H. Blaine, Jr. Calibrated geometries. *Acta Math.* **148** (1982), 47–157. (§6.5)

Hass, Joel, **Hutchings**, Michael, and **Schlafly**, Roger. The double bubble conjecture. *Elec. Res. Ann. AMS* **1** (1995), 98–102. (Preface, §14.0)

Hass, Joel, and **Schlafly**, Roger.

[1] Bubbles and double bubbles. *Amer. Scient.*, Sep.–Oct. 1996, 462–467. (§14.0)

[2] Double bubbles minimize. *Ann. Math.* **151** (2000), 459–515. (§14.0)

Heath, Sir Thomas. "A History of Greek Mathematics," Vol. II. Oxford University Press, London, 1921. (§15.1)

Heilmann, Cory, **Lai**, Yuan Y., **Reichardt**, Ben W., and **Spielman**, Anita. Component bounds for area-minimizing double bubbles, NSF "SMALL" undergraduate research Geometry Group report. Williams College, Williamstown, MA, 1999. (§§14.11–14.14)

Heintze, Ernst, and **Karcher**, Hermann. A general comparison theorem with applications to volume estimates for submanifolds. *Ann. Scient. Ec. Norm. Sup.* **11** (1978), 451–470. (§18.4)

Heppes, A.

[1] Isogonale sphärischen Netze. *Ann. Univ. Sci. Budapest Eötvös Sect. Math.* **7** (1964), 41–48. (§13.9)

[2] On surface-minimizing polyhedral decompositions. *Disc. Comp. Geom.* **13** (1995), 529–539. (§13.9)

Heppes, Aladár, and **Morgan**, Frank. Planar clusters and perimeter bounds. *Phil. Mag.* **85** (2005), 1333–1345. (§13.1)

Hildebrandt, Stefan. Free boundary problems for minimal surfaces and related questions. *Comm. Pure Appl. Math.* **39** (1986), *S*111–*S*138. (§12.3)

Hildebrandt, Stefan, and **Tromba**, Anthony. "The Parsimonious Universe." Copernicus, Springer-Verlag, New York, 1996. (Preface, Figure 16.4.1)

Hoffman, David. The computer-aided discovery of new embedded minimal surfaces. *Math. Intelligencer* **9** (1987), 8–21. (Figure 6.1.3a)

Hoffman, David, and **Meeks**, W. H., III. A complete embedded minimal surface in $\mathbf{R}^3$ with genus one and three ends. *J. Diff. Geom.* **21** (1985), 109–127. (Figure 6.1.3a)

Hoffman, David, and **Spruck**, Joel. Sobolev and isoperimetric inequalities for Riemannian submanifolds. *Comm. Pure Appl. Math.* **27** (1974), 715–727. Correction, **28** (1975), 765–766. (§17.3)

Hoffman, David, **Wei**, Fusheng, and **Karcher**, Hermann. Adding handles to the helicoid. *Bull. AMS* **29** (1993), 77–84. (Preface, Figure 6.1.3b)

Hopf, E. Elementare Bemerkungen über die Lösungen partieller Differentialgleichungen zweiter Ordnung vom elliptischen Typus, Sitzungberichte der Preussischen Akademie der Wissenshaften zu Berlin. *Phys.-Math. Klasse* **19** (1927), 147–152. (§§8.5, 10.4)

Howards, Hugh, **Hutchings**, Michael, and **Morgan**, Frank. The isoperimetric problem on surfaces. *Am. Math. Monthly* **106** (1999), 430–439. (§13.2)

Howe, Sean. The log-convex density conjecture and vertical surface area in warped products. *Adv. Geom.* **15** (2015), 455–468. (§§20.2, 20.5)

Hsiang, Wu-Yi. Talk at First MSJ International Research Institute, Sendai, 1993. (§13.2)

Hsiang, W.-T., and **Hsiang**, W.-Y. On the uniqueness of isoperimetric solutions and imbedded soap bubbles in noncompact symmetric spaces. *Invent. Math.* **85** (1989), 39–58. (§13.2)

Huisken, Gerhard. Flow by mean curvature of convex surfaces into spheres. *J. Diff. Geom.* **20** (1984), 237–266. (§16.7)

Hutchings, Michael. The structure of area-minimizing double bubbles. *J. Geom. Anal.* **7** (1997), 285–304. (§§13.0, 13.2, Chapter 14)

Hutchings, Michael, **Morgan**, Frank, **Ritoré**, Manuel, and **Ros**, Antonio. Proof of the Double Bubble Conjecture. *Ann. Math.* **155** (2002), 459–489. Research announcement ERA AMS **6** (2000), 45–49. (Preface, Chapter 14, §19.1)

Ilmanen, T. A strong maximum principle for singular minimal hypersurfaces. *Calc. Var. PDE* **4** (1996), 443–467. (§10.4)

Isenberg, Cyril. "The Science of Soap Films and Soap Bubbles." Dover, New York, 1992. (§13.9)

Jiang, Y., **Janiaud**, E., **Flament**, C., **Glazier**, J. A., and **Graner**, F. Energy landscape of 2D fluid foams with small area disparity. "Foams, Emulsions and Their Applications, Proceedings of the 3rd Euroconference on Foams, Emulsions and Their Applications (Delft, 4–8 June 2000)" (Pacelli Zitha, John Banhart, and Guy Vervist, Eds.), Verlag MIT, Bremen, 2000, 321–327. (§13.1)

Jordan, Cimille. Essais sur la geometrie a n dimensions. *Bull. Soc. Math. France* **3** (1875), 103–174. (§4.1)

Joyce, Dominick. Lectures on special Lagrangian geometry, "Global Theory of Minimal Surfaces" (Proc. Clay Research Institution 2001 Summer School, MSRI, David Hoffman, Ed.,), *Amer. Math. Soc.*, 2005. (§6.5)

Kanigel, Robert. Bubble, bubble: Jean Taylor and the mathematics of minimal surfaces. *The Sciences* (May/June 1993), 32. (§13.8)

Kinderleharer, D., **Nirenberg**, L., and **Spruck**, J. Regularity in elliptic free boundary problems, I. *J. Anal. Math.* **34** (1978), 86–119. (§13.9)

Klarreich, Erica G. Foams and honeycombs, *Am. Sci.* **88** (March/April 2000), 152–161. (§§15.0, 15.10)

Kleiner, Bruce. An isoperimetric comparison theorem. *Invent. Math.* **108** (1992), 37–47. (§17.3)

Knorr, Wilbur Richard. "The Ancient Tradition of Geometric Problems." Birkhäuser, Boston. (§13.2)

Kolesnikov, Alexander, and **Zhdanov**, Roman. On isoperimetric sets of radially symmetric measures. "Concentration, functional inequalities and isoperimetry." *Contemp. Math.* **545** (2011), 123–154. (§20.5)

Korevaar, Nicholas J., **Kusner**, Rob, and **Solomon**, Bruce. The structure of complete embedded surfaces with constant mean curvature. *J. Diff. Geom.* **30** (1989), 465–503. (§14.7)

Krantz, Steven G., and **Parks**, Harold R. "Geometric Integration Theory." Birkhäuser, Boston, 2008. (Preface)

Kusner, Robert B., and **Sullivan**, John M.

[1] Comparing the Weaire–Phelan equal-volume foam to Kelvin's foam. *Forma* **11** (1996), 233–242. Reprinted in Weaire [2]. (§15.10)

[2] Möbius energies for knots and links, surfaces and submanifolds. *AMS/IP Studies Adv. Math.* **2** (1997), Part 1, 570–604. (§16.6)

Lamarle, Ernest. Sur la stabilité des systèmes liquides en lames minces. *Mém. Acad. R. Belg.* **35** (1864), 3–104. (§13.9)

Lawlor, Gary.

[1] The angle criterion. *Invent. Math.* **95** (1989), 437–446. (§6.5)

[2] Proving area minimization by directed slicing. *Indiana U. Math. J.* **47** (1998), 1547–1592. (§6.5)

[3] A sufficient condition for a cone to be area-minimizing. *Mem. AMS* **91** (446) (1991). (§6.5)

Lawlor, Gary, and **Morgan**, Frank.

[1] Curvy slicing proves that triple junctions locally minimize area. *J. Diff. Geom.* **44** (1996), 514–528. (§§6.5, 13.9)

[2] Paired calibrations applied to soap films, immiscible fluids, and surfaces or networks minimizing other norms. *Pac. J. Math.* **166** (1994), 55–83. (§6.5)

Lawson, H. Blaine, Jr. "Lectures on Minimal Submanifolds," Vol. 1. Publish or Perish, 1980. (Preface, §§1.2, 8.1)

Lawson, H. Blaine, Jr., and **Osserman**, Robert. Non-existence, non-uniqueness and irregularity of solutions to the minimal surface system. *Acta Math.* **139** (1977), 1–17. (§6.2)

Le Hong, Van. Relative calibrations and the problem of stability of minimal surfaces. "Lecture Notes in Mathematics," No. 1453, pp. 245–262. Springer-Verlag, New York, 1990. (§6.5)

Ledoux, Michel, and **Talagrand**, Michel. "Probability in Banach Spaces." Springer-Verlag, New York, 2002. (§18.1)

Leonardi, Gian-Paolo. Infiltrations in immiscible fluids systems. *Proc. R. Soc. Edinburgh Sect. A* **131** (2001), 425–436. (§16.3)

Li, Yifei, **Mara**, Michael, **Rosa Plata**, Isamar, and **Wikner**, Elena. "Tiling with penalties and isoperimetry with density." Geometry Group report, Williams College, 2010. (Exercise 20.3)

Lichnerowicz, André. Variétés riemanniennes à tenseur C non négatif. *C. R. Acad. Sci. Paris Sér. A–B* **271** (1970), A650–A653.

Likos, C. N., and **Henley**, C. L. Complex alloy phases for binary hard-disc mixtures. *Phil. Mag. B* **68** (1993), 85–113. (§15.9)

Lopez, Robert, and **Baker**, Tracy Borawski. The double bubble problem on the cone. *NY J. Math.* **12** (2006), 157–167. (§19.0)

Lytchak, Alexander and **Wenger**, Stefan. Area minimizing discs in metric spaces. arXiv.org, 2015. (§1.2)

Mackenzie, D. See Nance.

MacPherson, Robert D., and **Srolovitz**, David J. The von Neumann relation generalized to coarsening of three-dimensional microstructures. *Nature* **446** (April 2007), 1053–1055. (§13.13)

Mandelbrot, Benoit B.
  [1] "The Fractal Geometry of Nature." Freeman, New York, 1983. (§2.3)
  [2] "Fractals." Freeman, San Francisco, 1977. (§2.3)

Masters, Joseph D. The perimeter-minimizing enclosure of two areas in S^2. *Real Anal. Exch.* **22** (1996/97), 645–654. (§§14.23, 19.4)

Maurmann, Quinn, and **Morgan**, Frank. Isoperimetric comparison theorems for manifolds with density. *Calc. Var. Partial Differential Equations* **36** (2009), 1–5. (§20.5)

McKean, H. P. Geometry of differential space. *Ann. Prob.* **1** (1973), 197–206. (§18.1)

Meeks, William H., III, and **Pérez**, Joaquín. The classical theory of minimal surfaces. *Bull. Amer. Math. Soc.* **48** (2011), 325–407. (§6.0)

Montesinos Amilibia, A. Existence and uniqueness of standard bubbles of given volumes in $\mathbf{R}^n$. *Asian J. Math.* **5** (2001), 25–32. (§14.2)

Morgan, Frank.

[1] Area-minimizing surfaces, faces of Grassmannians, and calibrations. *Am. Math. Monthly* **95** (1988), 813–822. (§§4.1, 6.5)

[2] Calibrations and new singularities in area-minimizing surfaces: a survey. "Variational Methods" (Proc. Conf. Paris, June 1988), (H. Berestycki, J.-M. Coron, and I. Ekeland, Eds.). *Prog. Nonlinear Diff. Eqns. Appl.* **4**, 329–342. Birkhäuser, Boston, 1990. (§6.5)

[3] Clusters minimizing area plus length of singular curves. *Math. Ann.* **299** (1994), 697–714. (§§13.0, 13.3, 13.11)

[4] "Compound Soap Bubbles, Shortest Networks, and Minimal Surfaces." AMS video, 1993. (§13.0)

[5] The Double Bubble Conjecture, *FOCUS*, Math. Assn. Amer. (December 1995). (§14.0)

[6] On finiteness of the number of stable minimal hypersurfaces with a fixed boundary. *Indiana Univ. Math. J.* **35** (1986), 779–833. (§6.1)

[7] Generic uniqueness results for hypersurfaces minimizing the integral of an elliptic integrand with constant coefficients. *Indiana Univ. Math. J.* **30** (1981), 29–45. (§8.5)

[8] The hexagonal honeycomb conjecture. *Trans. AMS* **351** (1999), 1753–1763. (§15.3)

[9] Immiscible fluid clusters in $\mathbf{R}^2$ and $\mathbf{R}^3$. *Mich. Math. J.* **45** (1998), 441–450. (§§16.2, 16.3)

[10] Lower-semicontinuity of energy of clusters. *Proc. R. Soc. Edinburgh* **127**A (1997), 819–822. (§16.4)

[11] $(\mathbf{M}, \varepsilon, \delta)$-minimal curve regularity. *Proc. AMS* **120** (1994), 677–686. (§§11.3, 13.10)

[12] Mathematicians, including undergraduates, look at soap bubbles. *Am. Math. Monthly* **101** (1994), 343–351. (§13.0)

[13] Minimal surfaces, crystals, shortest networks, and undergraduate research. *Math. Intelligencer* **14** (Summer, 1992), 37–44. (§13.0)

[14] Review of "Mathematics and Optimal Form" by S. Hildebrandt and A. Tromba. *Am. Math. Monthly* **95** (1988), 569–575. (Preface)

[15] Review of "The Parsimonious Universe" (Hildebrandt/Tromba). *Am. Math. Monthly* **104** (April, 1997), 377–380. (Preface, §15.8)

[16] "Riemannian Geometry: A Beginner's Guide." A. K. Peters, Wellesley, MA, 1998. (§13.2)

[17] Size-minimizing rectifiable currents. *Invent. Math.* **96** (1989), 333–348. (Gap in proof of Theorem 2.11 corrected by De Pauw and Hardt.) (§11.3)

[18] Soap bubbles and soap films. "Mathematical Vistas: New and Recent Publications in Mathematics from the New York Academy of Sciences" (J. Malkevitch and D. McCarthy, Eds.), Vol. 607. New York Academy of Sciences, New York, 1990. (§6.1)

[19] Soap bubbles in R^2 and in surfaces. *Pac. J. Math.* **165** (1994), 347–361. (§§13.1, 13.4, 13.10)

[20] Soap films and mathematics. *Proc. Symp. Pure Math.* **54** (1993), Part 1, 375–380. (§11.3)

[21] Soap films and problems without unique solutions. *Am. Sci.* **74** (1986), 232–236. (§6.1)

[22] Surfaces minimizing area plus length of singular curves. *Proc. AMS* **122** (1994), 1153–1164. (§13.11)

[23] Survey lectures on geometric measure theory. "Geometry and Global Analysis: Report of the First MSJ International Research Institute, July 12–13, 1993" (Takeshi Kotake, Seiki Nishikawa, and Richard Schoen, Eds.), 87–110. Tohoku University, Sendai, Japan, 1993. (§13.9)

[24] What is a surface? *Am. Math. Monthly* **103** (May, 1996), 369–376. (Preface, §1.2, Figure 4.3.3)

[25] Manifolds with density. *Notices AMS* **52** (2005), 853–858. (§§8.5, 18.0)

[26] Myers' theorem with density. *Kodai Math. J.* **29** (2006), 455–461. (§18.8)

[27] Regularity of isoperimetric hypersurfaces in Riemannian manifolds. *Trans. AMS* **355** (2003), 5041–5052. (§8.5)

[28] Existence of least-perimeter partitions. *Phil. Mag. Lett.* (2008). (§15.0)

[29] Manifolds with density http://sites.williams.edu/Morgan/2010/03/15/manifolds-with-density/ and Fuller references http://sites.williams.edu/Morgan/2010/03/16/manifolds-with-density-fuller-references/ (§§18.0, 20.0, 20.4)

[30] The Log-Convex Density Conjecture. http://sites.williams.edu/Morgan/2010/04/03/the-log-convex-density-conjecture (§20.2)

[31] Variation formulae for perimeter and volume densities. http://sites.williams.edu/Morgan/2010/06/22/variation-formulae-for-perimeter-and-volume-densities (§20.2)

Morgan, Frank, **French**, Christopher, and **Greenleaf**, Scott. Wulff clusters in R^2. *J. Geom. Anal.* **8** (1998), 97–115. (§16.5)

Morgan, Frank, and **Johnson**, David L. Some sharp isoperimetric theorems for Riemannian manifolds. *Indiana U. Math. J.* **49** (2000), 1017–1041. (§§13.2, 17.3)

Morgan, Frank, and **Pratelli**, Aldo. Existence of isoperimetric regions in R^n with density. *Ann. Global Anal. Geom.* **43** (2013), 331–365. (§20.4)

Morgan, Frank, and **Ritoré**, Manuel. Isoperimetric regions in cones. *Trans. AMS* **354** (2002), 2327–2339. (§18.10)

Morgan, Frank, and **Taylor**, Jean. The tetrahedral point junction is excluded if triple junctions have edge energy. *Scr. Metall. Mater.* **15** (1991), 1907–1910. (§13.11)

Morrey, C. B., Jr. "Multiple Integrals in the Calculus of Variations", Springer-Verlag, New York, 1966. (§1.2)

Murdoch, Timothy A. Twisted calibrations. *Trans. AMS* **328** (1991), 239–257. (§6.5)

Myers, S. B. Riemannian manifolds with positive mean curvature. *Duke Math. J.* **8** (1941), 401–404. (§18.8)

Nance [Mackenzie], Dana. Sufficient conditions for a pair of n-planes to be area-minimizing. *Math. Ann.* **279** (1987), 161–164. (§6.5)

Nitsche, Johannes C. C.

[1] The higher regularity of liquid edges in aggregates of minimal surfaces. *Nachr. Akad. Wiss. Göttingen Math.-Phys. Klasse* **2** (1977), 75–95. (§13.9)

[2] "Vorlesungen über Minimalflächen." Springer-Verlag, New York, 1975 [Translation: "Lectures on Minimal Surfaces." Cambridge University Press, New York, 1989]. (Preface, §§1.2, 8.1, 8.4)

Osserman, Robert.

[1] The isoperimetric inequality. *Bull. AMS* **84** (1978), 1182–1238. (§17.1)

[2] "A Survey of Minimal Surfaces." Dover, New York, 1986. (Preface, §§1.2, 8.1)

Pauling, Linus. "The Nature of the Chemical Bond and the Structure of Molecules and Crystals: an Introduction to Modern Structural Chemistry," 3rd ed. Cornell University Press, Ithaca, NY, 1960. (§15.12)

Pedrosa, Renato H. L. The isoperimetric problem in spherical cylinders. *Ann. Global Anal. Geom.* **26** (2004), 333–354. (§13.2)

Pedrosa, Renato H. L., and **Ritoré**, Manuel. Isoperimetric domains in the Riemannian product of a circle with a simply connected space form and applications to free boundary problems. *Indiana Univ. Math. J.* **48** (1999), 1357–1394. (§13.2)

Perelman, Grisha. The entropy formula for the Ricci flow and its geometric applications. arXiv.org (2002). (Chapter 18)

Peterson, Ivars.

[1] Constructing a stingy scaffolding for foam. *Science News* (March 5, 1994). (§15.0)

[2] The honeycomb conjecture. *Science News* **156** (July 24, 1999), 60–61. (§15.0)

Plateau, J. "Statique Expérimentale et Théorique des Liquides Sournis aux Seules Forces Moléculaires." Gauthier-Villars, Paris, 1873. (Preface, §13.8).

Preiss, David. Geometry of measure in $\mathbf{R}^n$: distribution, rectifiability, and densities. *Ann. Math.* **1** (1987), 537–643. (§3.12)

Qian, Zhongmin. Estimates for weighted volumes and applications. *Q. J. Math.* **48** (1997), 235–242. (§18.8)

Rado, Tibor. "On the Problem of Plateau." Springer-Verlag, New York, 1933 [reprinted 1971]. (§1.2)

Reichardt, Ben W. Proof of the double bubble conjecture in $\mathbf{R}^n$. *J. Geom. Anal.* **18** (2008), 172–191. (§§14.0, 14.21, 14.22, 19.2)

Reichardt, Ben W., **Heilmann**, Cory, **Lai**, Yuan, Y., and **Spielman**, Anita. Proof of the double bubble conjecture in $\mathbf{R}^4$ and certain higher dimensional cases. *Pac. J. Math.* **208** (2003), 347–366. (Preface, §14.0)

Reifenberg, E. R.

[1] Solution of the Plateau problem for m-dimensional surfaces of varying topological type. *Acta Math.* **104** (1960), 1–92. (§§1.0, 4.0, 13.9)

[2] An epiperimetric inequality related to the analyticity of minimal surfaces. *Ann. Math.* **80** (1964), 1–14. (§§1.0, 4.0, 13.9)

[3] On the analyticity of minimal surfaces. *Ann. Math.* **80** (1964), 15–21. (§§1.0, 4.0, 13.9)

[4] A problem on circles. *Math. Gazette* **32** (1948), 290–292. (§2.7)

Ritoré, Manuel. Applications of compactness results for harmonic maps to stable constant mean curvature surfaces. *Math. Z.* **226** (1997), 465–481. (§13.2)

Ritoré, Manuel, and **Ros**, Antonio.

[1] The spaces of index one minimal surfaces and stable constant mean curvature surfaces embedded in flat three manifolds. *Trans. AMS* **258** (1996), 391–410. (§13.2)

[2] Stable constant mean curvature tori and the isoperimetric problem in three space forms. *Comm. Math. Helv.* **67** (1992), 293–305. (§13.2)

Ros, Antonio.

[1] The isoperimetric problem. "Global Theory of Minimal Surfaces" (David Hoffman, Ed.). *Am. Math. Soc.*, Providence, RI, 2005, 175–209. (§§13.2, 18.2, 18.4)

[2] Stable periodic constant mean curvature surfaces and mesoscopic phase separation. *Interfaces Free Bound.* **9** (2007), 355–365. (§13.2)

Rosales, César, **Cañete**, Antonio, **Bayle**, Vincent, and **Morgan**, Frank. On the isoperimetric problem in Euclidean space with density. *Calc. Varn. PDE* **31** (2008), 27–46. (§18.9)

Ross, Sydney. Bubbles and foam. *Ind. Eng. Chem.* **61** (1969), 48–57. (§13.12)

Schmidt, Erhard. Beweis der isoperimetrischen Eigenschaft der Kugel im hyperbolischen und sphrärischen Raum jeder Dimensionenzahl. *Math. Z.* **49** (1943), 1–109. (§13.2)

Schoen, R., and **Simon**, L. Regularity of stable minimal hypersurfaces. *Comm. Pure Appl. Math.* **34** (1981), 741–797. (§8.5)

Schoen, R., **Simon**, L., and **Almgren**, F. J. Regularity and singularity estimates on hypersurfaces minimizing parametric elliptic variational integrals. *Acta Math.* **139** (1977), 217–265. (§8.5)

Schoen, R., **Simon**, L., and **Yau**, S.-T. Curvature estimates for minimal hypersurfaces. *Acta Math.* **134** (1975), 275–288. (§8.5)

Schoen, R., and **Yau**, S.-T. On the proof of the positive mass conjecture in General Relativity. *Comm. Math. Phys.* **65** (1979), 45–76. (Preface, §16.7)

Schwarz, H. A. Beweis des Satzes, dass die Kugel kleinere Oberfläche besitzt, als jeder andere Körper gleichen Volumens. *Nachrichten Königlichen Gesellschaft Wissenschaften Göttingen* (1884), 1–13. (§13.2)

Serrin, J. On the strong maximum principle for quasilinear second order differential inequalities. *J. Funct. Anal.* **5** (1970), 184–193. (§8.5)

Simon, Leon

[1] Cylindrical tangent cones and the singular set of minimal submanifolds. *J. Diff. Geom.* **38** (1993), 585–652. (§11.1)

[2] Existence of surfaces minimizing the Willmore functional. *Comm. Anal. Geom.* **1** (1993), 281–326. (§16.6)

[3] "Lectures on Geometric Measure Theory." *Proc. Centre Math. Anal. Austral. Nat. Univ.* **3** (1983). (Preface, §§4.3, 4.11, 9.1, 12.4)

[4] Survey lectures on minimal submanifolds. "Seminar on Minimal Submanifolds" (E. Bombieri, Ed.). Princeton University Press, Princeton, NJ, 1983. (Preface)

Simons, James. Minimal varieties in Riemannian manifolds. *Ann. Math.* **88** (1968), 62–105. (§§8.1, 10.5)

Smale, Nathan. Singular homologically area minimizing surfaces of codimension one in Riemannian manifolds. *Invent. Math.* **135** (1999), 145–183. (§13.2)

Solomon, Bruce, and **White**, Brian. A strong maximum principle for varifolds that are stationary with respect to even parametric elliptic functionals. *Indiana Univ. Math. J.* **38** (1989), 683–691. (§10.4)

Stewart, Ian. Circularly covering clathrin. *Nature* **351** (May 9, 1991), 103. (§15.12)

Stroock, Daniel W. "Probability Theory: an Analytic View." Cambridge University Press, New York, 1993. (§§18.1, 18.2)

Sudakov, V. N., and **Tsirel'son**, B. S. Extremal properties of half-spaces for spherically invariant measures. *J. Soviet Math.* (1978), 9–18. (§18.2)

Sullivan, John M.

[1] "The Optiverse" and other sphere eversions. Proc. ISAMA99, San Sabastián, Spain (June, 1999) and Proc. Bridges, Kansas (July, 1999). (§16.7)

[2] Sphere packings give an explicit bound for the Besicovitch covering theorem. *J. Geom. Anal.* **4** (1994), 219–231. (Figure 2.7.1)

Sullivan, John M., **Francis**, George, and **Levy**, Stuart. The Optiverse. VideoMath Festival at ICM '98. Springer, New York, 1998, 7-minute video. (§16.7)

Sullivan, John M., and **Morgan**, Frank (Eds.). Open problems in soap bubble geometry. *Int. J. Math.* **7** (1996), 833–842. (Preface, §§13.9, 14.21, 17.1)

Tarnai, Tibor. The observed form of coated vesicles and a mathematical covering problem. *J. Mol. Biol.* **218** (1991), 485–488. (§15.12)

Taylor, Jean E.

[1] Boundary regularity for solutions to various capillarity and free boundary problems. *Comm. PDE* **2** (1977), 323–357. (§12.3)

[2] Crystalline variational problems. *Bull. AMS* **84** (1987), 568–588. (§§12.5, 16.4)

[3] Regularity of the singular sets of two-dimensional area-minimizing flat chains modulo 3 in $\mathbf{R}^3$. *Invent. Math.* **22** (1973), 119–159. (§11.1)

[4] The structure of singularities in soap-bubble-like and soap-film-like minimal surfaces. *Ann. Math.* **103** (1976), 489–539. (§§11.3, 13.9)

Taylor, J. E., **Cahn**, J. W., and **Handwerker**, C. A. Geometric models of crystal growth. *Acta Metall. Mater.* **40** (1992), 1443–1474 (Overview No. 98–I). (§16.7)

Teixeira, P. I. C., **Graner**, F., and **Fortes**, M. A.

[1] Lower bounds for the surface energy of two-dimensional foams. *Eur. Phys. J. E* **9** (2002), 447–452. (§13.1)

[2] Mixing and sorting of bidisperse two-dimensional bubbles. *Eur. Phys. J. E* **9** (2002), 161–169. (§§13.1, 15.9)

Thompson, D'Arcy Wentworth. "On Growth and Form," abridged edition. Cambridge University Press, New York, 1969. (Figure 15.1.1)

Thomson, William (Lord Kelvin). On the homogeneous division of space. *Proc. R. Soc. London* **55** (1894), 1–16. (§15.10)

Varro, Marcus Terentius. "On Agriculture." The Loeb Classical Library. Harvard University Press, Cambridge, MA, 1934. (§15.1)

Vaz, M. F., **Cox**, S. J., and **Alonso**, M. D. Minimum energy configurations of small bidisperse bubble clusters. *J. Phys. Condensed Matter* **16** (2004), 4165–4175. (§13.1).

Vaz, M. F., and **Fortes**, M. A. Two-dimensional clusters of identical bubbles. *J. Phys. Condensed Matter* **13** (2001), 1395–1411. (§13.1)

Vaz, M. F., **Fortes**, M. A., and **Graner**, F. Surface energy of free clusters of bubbles: an estimation. *Phil. Mag. Lett.* **82** (2002), 575–579. (§13.1)

Villani, Cédric. Optimal transport, old and new (lecture notes for 2005 Saint-Flour course). *www.umpa.ens-lyon.fr/~cvillani/cv.html#publicationlist.* (§§18.3, 18.8)

Weaire, Denis.

[1] Froths, foams and heady geometry. *New Scientist* (May 21, 1994). (§15.10)

[2] (Ed.), "The Kelvin Problem. Foam Structures of Minimal Surface Area." Taylor & Francis, London, 1996.

Weaire, Denis, and **Hutzler**, Stefan. "Physics of Foams." Oxford University Press, New York, 1999. (§§13.0, 13.1)

Weaire, Denis, and **Phelan**, Robert. A counter-example to Kelvin's conjecture on minimal surfaces. *Phil. Mag. Lett.* **69** (1994), 107–110. (§§15.0, 15.10)

Wecht, Brian, **Barber**, Megan, and **Tice**, Jennifer. Double crystals. *Acta Cryst. Sect. A* **56**, 92–95. (§16.5)

Wei, Guofang, and **Wylie**, Will. Comparison geometry for the Bakry–Émery Ricci tensor. *J. Differential Geom.* **83** (2009), 377–405. (§18.1)

Weyl, Hermann. "Symmetry." Princeton University Press, Princeton, NJ, 1952; Princeton Sci. Lib. ed., 1989. (§15.1)

White, Brian.

[1] Existence of least-area mappings of N-dimensional domains. *Ann. Math* **118** (1983), 179–185. (§1.2)

[2] Existence of least-energy configurations of immiscible fluids. *J. Geom. Anal.* **6** (1996), 151–161. (§§16.2, 16.3)

[3] A new proof of the compactness theorem for integral currents. *Comm. Math. Helv.* **64** (1989), 207–220. (§§3.17, 4.12, 5.4)

[4] A new proof of Federer's structure theorem for k-dimensional subsets of R^N. *J. AMS* **11** (1998), 693–701. (§3.17)

[5] Regularity of area-minimizing hypersurfaces at boundaries with multiplicity. "Seminar on Minimal Submanifolds" (E. Bombieri, Ed.), pp. 293–301. Princeton University Press, Princeton, NJ, 1983. (§8.4)

[6] Regularity of the singular sets in immiscible fluid interfaces and solutions to other Plateau-type problems. *Proc. Centre Math. Anal. Austral. Nat. Univ.* **10** (1985), 244–249. (§16.3)

[7] A regularity theorem for minimizing hypersurfaces modulo p. *Proc. Symp. Pure Math.* **44** (1986), 413–427. (§§8.5, 11.1)

[8] The structure of minimizing hypersurfaces mod 4. *Invent. Math.* **53** (1979), 45–58. (§§11.1)

[9] In preparation. (§13.10)

Whitney, Hassler. "Geometric Integration Theory." Princeton University Press, Princeton, NJ, 1957. (§4.7)

Wichiramala, Wacharin. Proof of the planar triple bubble conjecture. *J. Reine Angew. Math.* **567** (2004), 1–49. (§§13.1, 14.24)

Wickramasekera, Neshan. A sharp strong maximum principle and a sharp unique continuation theorem for singular minimal hypersurfaces. *Calc. Var. Partial Differential Equations* **51** (2014), 799–812. (§10.4)

Williams, Robert. "The Geometrical Foundation of Natural Structure: A Source Book of Design." Dover, New York, 1979. (§15.12)

Willmore, T. J. "Riemannian Geometry." Clarendon Press, Oxford, 1993. (§16.6)

Wirtinger, W. Eine Determinantenidentität und ihre Anwendung auf analytische Gebilde und Hermitesche Massbestimmung. *Monatsh. Math. Phys.* **44** (1936), 343–365. (§6.5)

Wong, Yung-Chow. Differential geometry of Grassmann manifolds. *Proc. Natl. Acad. Sci. USA* **57** (1967), 589–594. (§4.1)

Wylie, William. Some curvature pinching results for Riemannian manifolds with density. *Proc. Amer. Math. Soc.* **144** (2016), 823–836. (§18.3)

Yau, Shing-Tung. Isoperimetric constants and the first eigenvalue of a compact Riemannian manifold. *Ann. Scient. Éc. Norm. Sup.* **8** (1975), 487–507. (§17.3)

Young, L. C.

[1] On generalized surfaces of finite topological types. *Mem. AMS* No. 17 (1955), 1–63. (§§1.0, 4.0)

[2] Surfaces paramétriques generalisées. *Bull. Soc. Math. France* **79** (1951), 59–84. (§§1.0, 4.0)

Zhang, Qiancheng, **Yang**, Xiaohu, **Li**, Peng, **Huang**, Guoyou, **Feng**, Shangsheng, **Shen**, Cheng, **Han**, Bin, **Zhang**, Xiaohui, **Jin**, Feng, **Xu**, Feng, and **Lu**, Tian Jian. Bioinspired engineering of honeycomb structure – Using nature to inspire human innovation. *Prog. Mater. Sci.* **74** (2015), 332–400. (§13.0)

Zworski, Maciej. Decomposition of normal currents. *Proc. AMS* **102** (1988), 831–839. (§4.5)

Index of Symbols

Name Index

Subject Index

On my way.

Photograph courtesy of the Morgan family;
taken by the author's grandfather, Dr. Charles W. Selemeyer.

Printed in the United States
By Bookmasters